Exploring Lake
A Field

By the same author:

    Exploring Lakeland Tarns (1993)

COVER PICTURE. View through the "knick" where a waterfall is born and the Littledale stream drops to become the Scope Beck.

Exploring
LAKELAND WATERFALLS

A Field Guide

DON BLAIR

Lakeland Manor
Press

Published by Lakeland Manor Press,
Dunedin, 15 Manor Park, Keswick,
Cumbria CA12 4AB

First Published 2002

Text and photographs copyright © Don Blair 2002

All Rights Reserved. No part of this publication may be reproduced, stored in a retrieval system, transmitted in any form or by any means - electronic, mechanical, photocopying, recording, or otherwise - without prior written permission from the publisher.

Printed and bound in Great Britain by Athenaeum Press Ltd., Gateshead, Tyne & Wear.

British Library Cataloguing Data
A catalogue record for this book is available
from the British Library.
ISBN 0 9543904 0 7

Contents

	Pages
Introduction:	7 - 13
Sector 1: Falls in Upper Derwent and Borrowdale.	14 - 38
Sector 2: Falls Back o'Skiddaw and Caldbeck Fells.	39 - 49
Sector 3: Falls around Thirlmere and St John's.	50 - 69
Sector 4: Falls in the Ullswater Valley.	70 - 86
Sector 5: Falls in Haweswater and Swindale Valleys.	87 - 97
Sector 6: Longsleddale and Kentmere.	98 - 105
Sector 7: Falls in the Windermere Catchment.	106 - 128
Sector 8: Falls in the Coniston Catchment.	120 - 141
Sector 9: Dunnerdale.	142 - 150
Sector 10: Falls in the Esk Valley.	151 - 163
Sectors 11 and 12: Falls in Wasdale and Ennerdale.	164 - 171
Sector 13: Falls around Buttermere and Loweswater.	172 - 183
Index:	184 - 188

Acknowledgements

In producing a publication of localised interest such as this, it is to family and friends one naturally turns for support and assistance or to local people for advice. As regards the former, it is my wife Gladys I have to thank for converting illegible scrawl into readable typescript as well as domestic support to keep the author functioning. For technical support I owe much to my stepson Neil as well as the legwork for the final fall when the author's own limb fell apart. To these two I express my gratitude for enabling the project to reach its conclusion.

It was the insistence of Silvia Pilling that initiated the project in the first place and to whom I apologise for the long delay. My appreciation too, to Joyce Wilson author of the "Muck---" farm stories who solved the problem of finding a publisher and unwittingly created even further problems! Thanks Joyce for all the help and advice.

Finally I also acknowledge the information regarding the gorge at Tilberthwaite supplied by Phil Clague of the LDNPA Ranger Service.

NOTE. The details of routes given in this guide do not imply a right of way although every effort has been made to ensure that the information presented in this book is accurate at the time of going to print. Readers are reminded here and elsewhere in the text that waterfalls are dangerous places, particularly where children are involved. The author/publisher cannot take responsibility for any accident or injury sustained on any of the walks. Appropriate clothing and footwear for hill walking should be worn and weather conditions checked before setting out.

The sketch maps are included to illustrate the distribution of waterfalls in each sector and are not intended to replace the appropriate Ordnance Survey Maps.

Introduction

Viewed in the World context the waterfalls of the English Lake District are infinitesimal; even Scotland yields higher and Wales can equal. Some obvious reasons for this lie in the altitude of the highlands in continental regions in which waterfalls occur and the vastness of the water-gathering areas or catchments where water gathers to power the falls.

The highest upland areas in Lakeland do not exceed 978 metres or 3210 feet as in Scafell Pike; as the area is divided into a series of relatively narrow ridges separated by deep valleys, there is no extensive upland area in which a reservoir of water may gather in order to sustain a vast and extended flow. By the time the precipitation collects in quantity it is already in the lakes in the valley bottoms where there are no great drops left to fall down! Plenty of rainfall yes, but it rapidly runs off the mainly impervious rocks of the fells and in doing so may create a short-lived, localised display of ferocity. If it can be caught in one of these rare moods, be fearful and cautious but it is against normal human behaviour to deliberately subject oneself to the conditions which bring about such phenomenon. In simple terms, we avoid going on the fells in exceptionally heavy rainfall. Only once did the writer encounter the "tail-end" of extreme conditions; an interesting and sobering experience. (See Launchy Gill).

If Lake District falls cannot equal in height or volume those of the great cataracts of Continental Europe, Asia, Africa, or North and South America, they do possess their own charms of setting and accessibility.

WHAT CREATES A WATERFALL ?

Fairly obviously a fall occurs where water, flowing as a stream, reaches a point where there is a sudden change of land level; the more abrupt the change or the greater the change of level, the more impressive the fall. There are several reasons for such changes of level, such as the physical uplift of land to form a plateau or the subsidence of a section of the land surface to form a rift valley, events common in our Planet's early history. Such dramatic earth movements in Africa and America have produced the World's greatest waterfalls: the Victoria Falls on the River Zambeze and the Kalamba Falls near Lake Tanganyika and in South America the Angel Falls, Earth's highest at 807m (2650 ft) first spotted by a Mr. Angel who was flying his aeroplane nearby.

As far as Lakeland is concerned it was the last Ice Age, which faded some 12,000 years ago, that was responsible for creating abrupt changes of land level, resulting in this region's larger falls. During that period when more snow fell in winter than melted in summer, former river valleys which one may assume had once existed among Cumbria's pre-glacial mountains, became choked with glacier ice which scoured and deepened them. When the great slow thaw came with climate change, the original valleys had been over-deepened into the typical "U" cross-profile where the Sourmilk Gill type of fall now descends from so-called "hanging" valleys left poised far above their main counterparts. Hanging valleys are reckoned to have been smaller tributaries of the main glacier, where present-day streams flow from Gillercomb in Borrowdale or tarn-bearing cirques such as Bleaberry Tarn in Buttermere, and Easedale Beck near Grasmere, each with its own Sourmilk Gill type of waterfall, usually a series of cascades.

Less dramatic falls occur on the streams that course along the deep valley bottoms. Much of Lakeland is formed of rocks of differing hardness such as the Borrowdale Volcanic Group of the central area and the Silurian slates and sandstones of the south. In the north the Skiddaw Slates are generally more homogeneous but intrusions of a volcanic nature introduced reefs of harder rock. When a stream crosses a band of hard rock the erosion of the bed at that point and upstream will be resisted but if downstream there are softer, more easily eroded rocks these will be worn away as far back as the band of hard rock. This will result in a step in the

Introduction.

stream bed (technically referred to as a **knick point** in the river profile) and, depending upon the abruptness of the change of level, either rapids, cascades or a waterfall may occur. A good example of this is Mill Fall on the Torver Beck. As through time the reef is worn back upstream it may leave in its wake a gorge or ravine. Even unintentional man-made waterfalls sometimes occur; one which comes to mind is at Banishead Quarry, near Torver.

SOME WATERFALL DEFINITIONS USED IN THE TEXT.

The term **waterfall** is used loosely to describe a near perpendicular descent of water, whether it be **unsupported,** where the water drops from lip to base without contact with the **back wall**, to one where the contact between water and back wall is maintained for all or most of the descent. Few Lakeland waterfalls are totally unsupported

Cascade. This describes descending water that is in contact with its bed or supporting rock for most of its descent, particularly where its smooth passage is broken by irregularities to produce a severely disturbed flow as it descends in leaps and bounds with lots of foam.

Water chute or **slide.** Water descending at an angle, in a smooth flow supported during its slanting passage by the underlying rock, like the kind of fun facility one might find at a leisure pool.

Spout. Water ejected under pressure through a gap (between boulders perhaps) to form a jet that projects outwards and down clear of the supporting rock.

Cataract. A voluminous flow of water consisting of one or a series of cascades.

Apron (broad), Ribbon (narrow) and Curtain falls. These are terms that loosely indicate the breadth of a waterfall. A curtain usually specifies a fall which descends without contact with the back wall and is "unsupported".

Plunge Pool. Most waterfalls drop into a hollow eroded at the

Exploring Lakeland Waterfalls.

foot or base of the fall or formed from rock fragments carried down in spate, to form a pool.

THE EFFECT OF RAINFALL.

It is axiomatic that the more prolific the rainfall, the more water the falls will carry and the more dramatic they become; in dry weather some will be reduced to a trickle. In very wet conditions falls will recur in certain regular channels; these are referred to here as Periodic Falls; one such is the White Lady Fall near Coniston. At such times fellsides may become laced with streams of white water when every beck becomes a cataract and every gill a cascade. At times like this the run-off from some falls may become destructive when the burden carried down causes road blockages as sometimes happens on the A591 beside Thirlmere. It is generally true to say that waterfalls are at their best during or soon after periods of heavy rainfall!

CONVENTIONS.

The terms "left" and "right" are relative terms and true only for an individual in accordance with the direction in which they are moving or facing at any one moment. A flow of water on the other hand may only move in one direction: downhill. The side of the flow on which one stands is tied to a convention related to flow. Face downstream or the same direction in which the water is running and the left side is the true left bank; the other side is the true right bank. This convention is used in the text. So left and right are always relative to the direction of flow.

SAFETY.

Waterfalls are dangerous places as writers of fiction are aware and as effective a natural feature for disposing of an unwanted character as a bog, a cliff or an avalanche. Although the Reichenbach Falls claimed Sherlock Holmes along with his villainous brother Moriarty in *"The Final Problem"*, it was only

Introduction.

readers' pressure that caused him to have miraculously survived in a later story. In real life alas, one seldom survives the physical injury or drowning associated with waterfall accidents; these have claimed several lives in the Lake District during recent years. This certainly applies to falls such as Moss Force on Newlands Hause, Dungeon Ghyll and the face of Force Crag, Braithwaite. Particularly notorious is Sourmilk Gill, Buttermere; even the modest Glenridding Beck has claimed its victim.

Waterfalls are wet areas and have steep drops, a fatal combination, not just at the lip itself but at the side of the ravine into which the fall plunges, often edged by a slippery, grassy slope convex in form that conceals the view within. A slip initiated while trying to obtain a better view is unlikely to be corrected before one hits verticality! Children, driven by curiosity that exceeds caution are particularly vulnerable. Remember that mobile phones are unlikely to operate from the bottom of a gorge so that a lone injured person may have a long wait before being rescued, even if they have survived the fall. The record is still held by a Mr. Crump who in 1921 survived for eighteen days in Pier's Gill, a fearsome gorge just off the Corridor Route to Scafell Pike. To explore the gill and the falls within is a mountaineering expedition and quite outside the scope of this book. (*A good account of this event and similar adventures may be found in A.H. Griffin ,"Adventuring in Lakeland"; Robert Hale, London 1980.*)

SIZE AND DISTANCE

In order to give some idea of scale and comparisons between falls, the heights of most have been suggested . The procedure of using surveying instruments or plumbing the drops with a weighted line did not appeal so estimates by eye were made; these probably became more accurate with practice. In accord with current trends, measurements were judged in metres (m) but as these are an approximation, they may be equally acceptable if also regarded as yards (or feet if multiplied by three). Walking distances are given in kilometres (k or km) and sometimes in miles for longer distances. Some indication of time is given for the outward journey only and does not include the return unless it is a "round trip". Quality time for viewing or refreshments is not included.

FINDING THE FALLS

The river drainage pattern within the Lake District National Park, is often described as radial in that it fans outward from some central point as if the Lakeland mountains had once been shaped like an inverted basin or dome so that rainfall had drained off in all directions, eventually eroding valleys that fanned outwards like the spokes of a wheel.

In an effort to be methodical the main fall-bearing streams have been grouped together according to the main valley into which each empties its water. There are twelve of these main radial valleys (Kentmere and Longsleddale I have counted as one), in addition to the small block (sector 2) to the north of the Skiddaw, Blencathra massif, referred to here as Back o'Skiddaw or Caldbeck Fells, making thirteen sectors.

The Ordnance Survey (OS) Maps used to cover the whole area are the Outdoor Leisure 1:25000, numbers 4 to 7 - but recently reissued as "Explorer" OL - (the new double- sided 4 and 5 give full coverage northwards). National Grid References are used in this book to aid location of the falls or streams on which they occur. *The sketch maps in the text are also for this purpose and are not intended to replace the OS maps.*

Only the major, interesting falls are included, or those that appealed at the time; that leaves scope for further exploration and who knows - out there somewhere may be that really, really big waterfall that still awaits discovery.

Introduction.

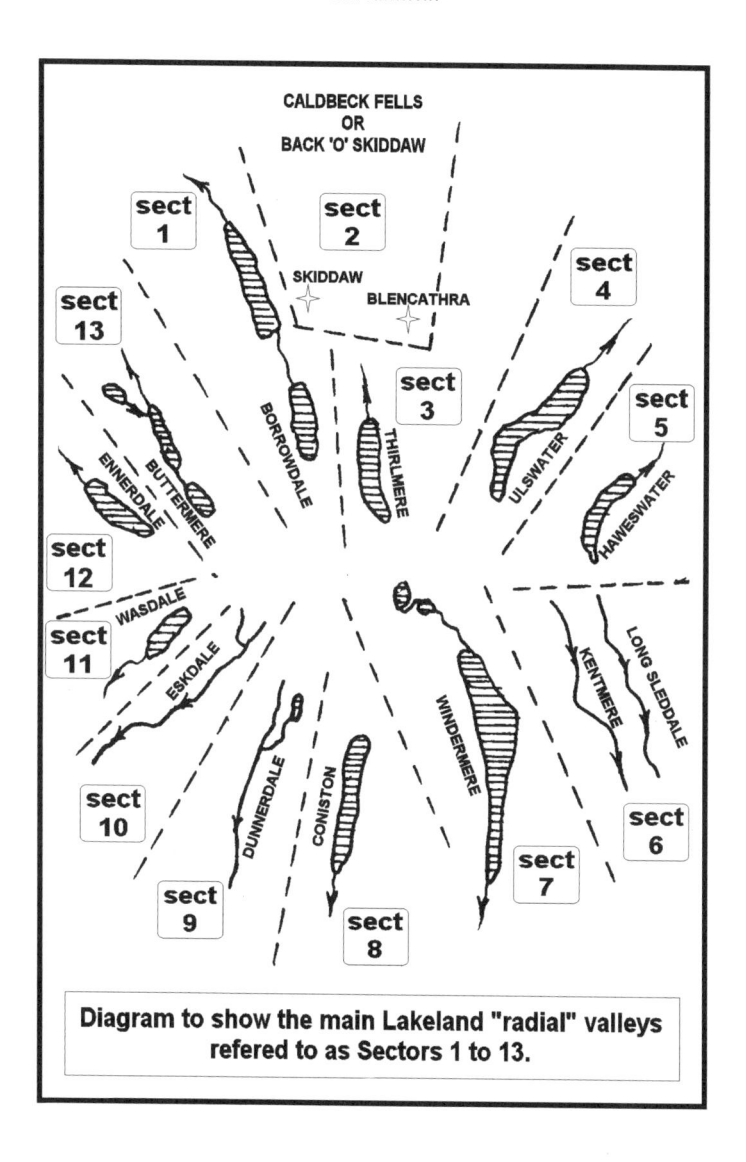

Diagram to show the main Lakeland "radial" valleys refered to as Sectors 1 to 13.

Sector One:

Falls in Upper Derwent and Borrowdale.

To begin near home seemed a sensible idea. It is also where many of the early visitors to the Lake District came to experience the Gothic horrors of the mountains and the dreadful downfall of Lodore.

In this sector will be found the principal falls of Borrowdale and its contributory valleys; also included are some streams that enter Bassenthwaite and for good measure, a couple of tributaries of the Glenderamackin under the heading Upper Derwent. The falls, or the streams on which falls may be found are as follows.

Falls index:-
(1) Gill Falls.
(2) Ashness Falls.
(3) Barrow Falls.
(4) Lodore Falls.
(5) Galleny Force, Stonethwaite Gill.
(6) Greenup Gill - falls.
(7) Langstrath - Blackmoss Pot..
(8) Comb Gill - fall on.
(9) Sourmilk Gill.
(10) Taylorgill Force.
(11) Scaleclose Force.
(12) Moss Force.
(13) Scope Beck - fall on.
(14) Newlands Beck - fall on.
(15) Ellers Beck - fall above.
(16) Low and High Force - Coledale Beck.
(17) Wythop Beck - fall on.
(18) Beckstones Gill - falls on.
(19) Slades Beck - falls on.
(20) Kilhow Beck and Gategill Beck - falls on.

Sector One: Falls in Upper Derwent and Borrowdale.

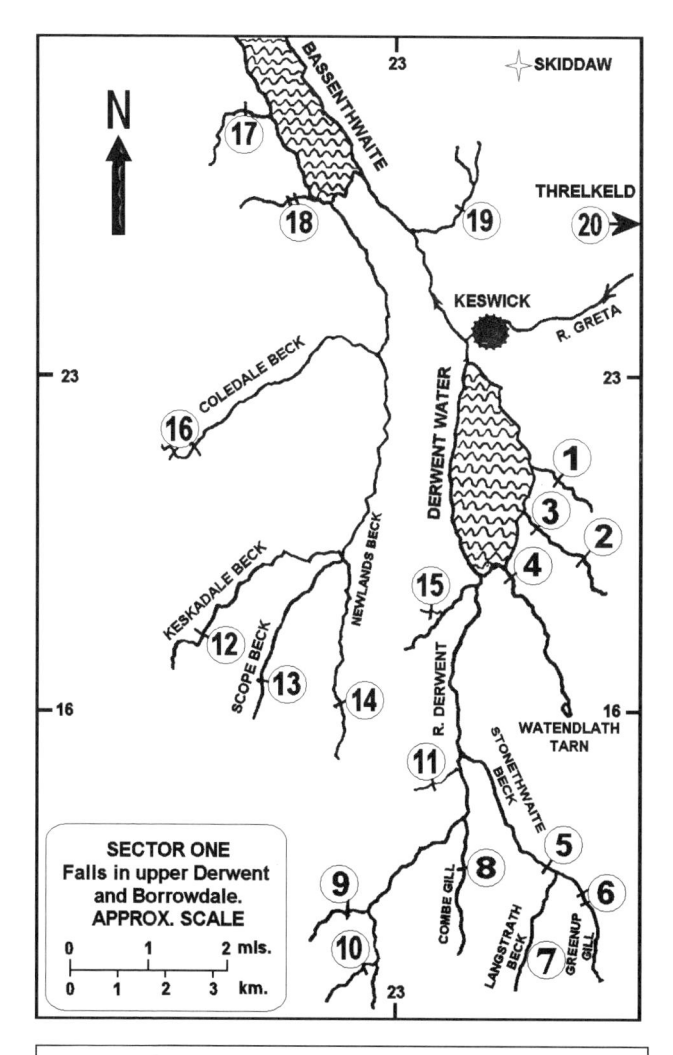

Sketch map to show the location of falls (1) to (20) in Sector One.

Exploring Lakeland Waterfalls.

(1) FALLS ON CAT GILL

Borrowdale.   **NY 275208.**

The National Trust pay and display car park at the northern end of Great Wood beside the Borrowdale Road at 273216 offers a convenient starting point for this, the most northerly fall in the valley. Cat Gill is not the best of them, being largely inaccessible for all but the hardy gill scrambler, dangerous to attempt to view from above and well hidden by trees, but the repaired path which ascends beside the true right side of the gill offers a short, sharp climb to the summit of Walla Crag.

From the Great Wood car park walk south through the wood to the gate and bear left until the descending Cat Gill is reached at a wooden footbridge. Ascend the steep path beside the gill, where it bears left the gill disappears within its cleft, hidden below the convex, grassy slope within its tree-choked ravine; but the roar from within tells of active water just out of sight. A safe and accessible viewpoint - aided by a pair of binoculars - is from the Falcon Crag side of the gill, of which more later.

Forget the fall for the moment and continue up the steep, gravelly path beside the wall until the ascent levels out and the stile is reached. Cross and walk up to the rocky summit of Walla Crag to regain breath and enjoy the fabulous view of Derwentwater and its islands, the town of Keswick backed by Skiddaw (931m or 3035 ft.) and the length of Bassenthwaite extending into the distance.

Return to the stile, re-cross the wall and take the well-trodden track that heads east away from the wall and the former steep ascent. In about five minutes cross the beck which is the main feeder stream for Cat Gill. A few metres downstream a pretty 6m. cascade trickles down rough rocks. Continue, with views to the right over Derwentwater, to the muddy right hand bend, beyond which the track rises slightly then resumes its southwards course; but here a faint path continues straight on through bracken to a

Sector One: Falls in Upper Derwent and Borrowdale.

greyish rock platform overlooking Cat Gill. Dodge around until a clear view of the 4m - 5m cascade, curved like a scimitar, with another small fall below can be seen; best visited when the trees in the gill are bare.

Return to the main path and continue to Ashness Bridge. Note that on the OS map the path from Walla Crag summit (379m.) to Ashness Bridge is more accurately indicated by the black pecked line, not the green one.

Great Wood to Walla Crag:- 1½k. (300m. ascent).   *Time:- 50 mims.*
Walla Crag to Ashness Bridge:- 3½k.   *Time:- 45 mins.*

(2) ASHNESS FALLS
Borrowdale.   **Barrow Beck.**   **NY 278194.**

A Lakeland calendar without a view of Ashness Bridge is a rarity; it is certainly photogenic. The Barrow Beck flows beneath and empties into Derwentwater. The bridge carries the road south to the tiny picturesque hamlet of Watendlath, the fictional home of Judith Paris in the Hugh Walpole novels.

There are some small lay-bys beyond the bridge with a short walk to Surprise View overlooking Derwentwater. From the bridge walk a few metres south along the road and take the signed path on the left that follows the true left bank of the Barrow Beck and which later becomes Ashness Gill beyond the fall. Numerous noisy cascades and a magnificent twin water slide can be seen on the way. Some 50m beyond the latched gate at the intake wall, leave the ascending path and take the lesser track that descends towards the beck. An awkward slippery scramble leads to the base of the fall. See the water cream over a sky-lined ledge some 30m above, to plunge towards one in a series of short falls and cascades. From this damp viewpoint return to the main track perhaps to continue uphill across the peaty moor to High Seat or return to Ashness Bridge to view the third fall of this group.

Ashness Bridge to Ashness Falls:- 1k.   *Time:- 30 mins.*

Exploring Lakeland Waterfalls.

(3) BARROW FALLS

Borrowdale.   **Barrow Beck.**   **NY 269199.**

After flowing beneath the famous Ashness Bridge, the Barrow Beck descends to the lake by way of Barrow Falls. These occur in the grounds and to the rear of Barrow House, the Derwentwater Youth Hostel. Courtesy suggests that non-members of the YHA should contact the warden for permission to view. (Tel.017687 77396). Front access is from the main Borrowdale Road, rear entry may be made from the Watendlath Road.

From Ashness Bridge walk northwards and downhill along the metalled road, past the Memorial Cairn to the memory of Bob Graham of the Lakeland 24 hour round (a long and strenuous walk over 42 peaks), to the gate marked "YHA Private Grounds". A steep track leads downhill to a series of slippery slate steps beside the cascading water to a little terrace at the base of the lower fall and in full view of the rear of the Hostel.

This is a three stage fall where water laces down steep rocks, the lower drop being of 6-7m, the upper pair combined being of similar magnitude. At the bottom, branches of berried rowan hung above the lower plunge pool from where the water curled away through the grounds to form a small delta in the lake at Barrow Bay.

(4) LODORE FALLS

Borrowdale.   **Watendlath Beck.**   **NY 265187.**

These falls are perhaps one of the best known in the Northern Lakes, long held in regard by early visitors to the dread mountains and awful storms of the region since the 18th Century. Although by no means dramatic in size or form compared with many Lakeland falls, it has the advantage of accessibility by being just 200m from the valley road , thus being well within the scope of the more adventurous Victorians

The great cleft of Lodore, rent between the towering cliffs of

Sector One: Falls in Upper Derwent and Borrowdale.

Shepherd's Crag and Gowder Crag by the Watendlath Beck, is filled with a series of short, steep rock steps and choked with massive boulders, the descending water falling over, around and between in a series of falls and cascades, dividing and crossing from side to side. Only the most severe conditions produce a sufficient head of water to cause it to o'erleap the impeding boulders and spread the width of the gorge as an expanding fan.

How the water comes down at Lodore.

It was probably on one of these rare occasions that Robert Southey was inspired to pen his description of "How does the water come down at Lodore?"

These falls lie to the rear of the Lodore Hotel, where for many years a small gate at the south side of the Hotel enabled access "To the Falls" for a small charge received by an honesty box. The nearest parking is at Kettlewell (NT) Car Park (267195) beside the lake; the Keswick launch also calls at the landing stage near the Hotel. Across the road from this car park a signed path directs walkers through Lodore Wood along muddy paths protected from traffic by the wall, southwards towards the fall. On the way, the track passes through a fence and back again by latched gates until one reaches the rear of the Lodore Hotel. Before this, the tumbling waters may be heard, there being a steady flow at all seasons.

An ascent of the falls may be made on the left side of the cleft, or the true right bank. Do not attempt to scale the scuffed slope straight ahead but zigzag on the graded path to view the pools and cascades in the wooded cleft below. From here, where the ascent eases, take the left path to a fence and stile then choose paths trending NE which will lead eventually to the tarmacked Ashness to Watendlath road. then left, or north will bring one to Ashness Bridge and the main Borrowdale Road where the pleasant lakeside path back southwards will bring one to the car park at Kettlewell.

Kettlewell to Lodore Falls:- under 1k.   *Time:- 30 mins.*
Lodore Falls to Kettlewell via Ashness Bridge:- 4k.   *Time:- 1¼ hrs.*

(5) GALLENY FORCE
Borrowdale.   **Stonethwaite Beck.**   **NY 273131.**

Galleny Force, little more than 2m in height, shoulders through a metre wide gap into an elongated basin of deep, clear water and round, grey boulders. Boughs hang above this placid pool where low rock walls, scoured smooth, suggest more active outbursts.

Sector One: Falls in Upper Derwent and Borrowdale.

This is a very modest fall to merit recognition by name on the OS map suggesting that ease of access and its attractive setting would make it a pleasant venue for a family stroll in less adventurous days.

The fall occurs at a sharp right-angle bend just downstream of the confluence of Greenup Gill and Langstrath Beck, uniting here as Stonethwaite Beck. The combination of two strong streams ensures a steady flow in all seasons. The approach from Stonethwaite, where there is some verge parking, is by a signed track to Langstrath which passes through a campsite and follows the left bank upstream to the fall.

From Langstrath Country Inn (hotel) :- 1½k.   *Time:- 30 mins.*

Galleny Force, Stonethwaite, may be reached with level walking.

(6) FALLS IN GREENUP GILL

Borrowdale.   **Stonethwaite.**   **NY 280123.**

Follow the path upstream beyond Galleny Force to the confluence and bear right along the left bank of the Langstrath Beck to the first footbridge. Cross and return downstream to Greenup Gill where it may be crossed by another footbridge. Walk upstream beside the beck on the clear, stony path that would lead eventually to Grasmere. Eagle Crag and Heron Crag rise impressively on the other side of the beck.

Pass the end of the first wall and through the gap in the next; as one approaches the third a 4m fall is visible below in a gloomy, shaded gorge. Continue through the gate and sheep pen and for some 150m beyond to where there is a tree-shaded pair of apron falls plunging side by side for 5m in two vertical columns, divided by a rock bastion.

Upstream, the gill in spate foams downhill towards one in a lather of white water, impressive in its combination of cascade, fall and water-slide, interspersed by calming pools.

One may return downhill by the same path but do not re-cross the bridge at the confluence but continue downstream on the right bank of the Stonethwaite Beck, over the Willygrass Gill (the outflow from Dock Tarn) to the Stonethwaite Bridge and cross to the village.

Stonethwaite to falls and return:- 5½k. (3½ mls.)   *Time:- 2 hrs.*

(7) LANGSTRATH BECK

Borrowdale.   **Blackmoss Pot.**   **NY 267113.**

There are no falls of merit here; the only feature of interest is Blackmoss Pot, a well-known local swimming pool where the beck takes the form of a long, deep pool within a sheer-sided ravine, so look before you leap.

(8) FALLS ON COMB GILL

Borrowdale.   **NY 253132.**

There are several means of access to this modest fall and associated cascades, the easiest being by a recently opened

Sector One: Falls in Upper Derwent and Borrowdale.

permissive section of path from Stonethwaite, where there is convenient parking at the verges of the valley road. The alternative approach is from the car park at Seatoller. The Borrowdale bus from Keswick will also provide convenient transport to Stonethwaite road end. From here, the steady uphill walk of about one mile or 3k will bring one to the fall. From the little Church at 258140, walk through the farmyard and on to the field path with the wall on the left hand. Shortly, a prominent sign indicates the right hand track towards the Borrowdale Road or straight on to the new permissive section of path. At the open gateway to the next field, stop. Ahead is a wooden shed but on the left hand is a waymark, easily missed, and a swing gate. Pass through and along the wet path with the wall now on the right hand. The next gate is shared with a tiny beck which approaches from ahead then sneaks through a hole in the right hand wall. Continue on the same line to the next waymark which directs one slantwise left and uphill to a prominent posted sign where a well-worn path may be followed to the fall, then to the intake wall where a gateway gives access to the Comb Gill Valley proper.

At the sign post, the alternative route from the Borrowdale Road joins the path. This starts from opposite the northern end of Mountain View Cottages and crosses Comb Gill by the little stone bridge where there is another sign directing walkers straight up the fellside.

The main fall is reached first but it may be necessary to descend from the path to a safe viewpoint just below the small, but dangerous ravine. Above the fall a clutter of boulders and projecting rocks cause the gill to divide into a number of ragged streams but all gather for the final plunge into the sheer-sided declivity, first in two short steps then the long ribbon fall of 6-7m into an oval pool shaded by trees and edged by fern where the high summer sun glances down. The pool's overspill slants away in a ruffled water slide to the downstream weir and water treatment works. Upstream, at intervals two pretty cascades fan

Exploring Lakeland Waterfalls.

down rounded breasts of rock. Across the gill is the main Comb Gill track which begins opposite the southern end of Mountain View Cottages and is the more popular route into the comb for those heading onwards towards Tarn at Leaves, Raven Crag or Glaramara.

Stonethwaite Church to fall:- 1½k.   *Time:- 35 mins.*

(9) SOURMILK GILL
Borrowdale.	Seathwaite.	NY 226123.

This is one of several of that name; there is another in Buttermere and one that flows out of Easedale Tarn, near Grasmere. The name is picturesquely descriptive of the impression given when in spate. There is usually no difficulty in spotting the force as it descends the fellside above Seathwaite Farm, even when partially concealed by summer greenery. Generations of walkers have parked their cars, angled to the wall beside the road and below the farm, for this is the launching pad for so many of the high fells; the early driver gets the parking space. Enter the cobbled farmyard and dive through the signed

The sparkling top fall on Sourmilk Gill.

Sector One: Falls in Upper Derwent and Borrowdale.

tunnel on the right between the barns and cross the bridge; the path left leads to Taylorgill Force, a later objective. Head up the very steep fell on the restructured path first in zigzags, then up steep water-worn rock, an easy scramble. The lower section of the force is tree-shrouded but can certainly be detected by ear; a little exploration to the right of the graded path will reveal cascades extending for about 12m, where the water bounces over short projecting rocks in its noisy descent.

Above the trees another section of cascades, observable from the track, extends for some 15m and consists of a chain of merry cascades, falls and water slides, this section being that which is so prominent from the farm below. As one proceeds uphill, there are several smaller falls and cascades for the adventurous to discover. Eventually the steep track relents as one approaches the intake wall and gate to the open fell.

But if you think that's the end you would be very wrong! The gem that most have missed is still to come. Continue uphill on the track for 100m; as it angles left your ears may inform you of falling water to the right, for partially concealed behind a rocky bluff is a fine free fall of some 5m where the stream that drains the shallow, tarnless cirque of Gillercomb first dives into the gill. The descending water is split by two short pinnacles and shattered into a milliard sparkling diamonds when caught by the morning sunlight, the veritable jewel in the crown of S.M.Gill.

300m. (1000ft.) climb from Seathwaite Farm:- as long as it takes.

From here, with the ascent behind one, the way is now open to take the well-used path in a leftwards, southerly direction, to Green Gable then the descent to Windy Gap. From here, further choices must be made - either an ascent of Great Gable or a return down Aaron Slack to Styhead Tarn and back to Seathwaite.

(10) TAYLORGILL FORCE
Borrowdale.   Styhead Gill.   **NY 230110.**

Styhead Gill drains from its mother tarn of that name and flows to

Exploring Lakeland Waterfalls.

join Grains Gill just downstream of Stockley Bridge; united, they form the infant R. Derwent. Just before the confluence, Styhead Gill enters a restricted ravine below Base Brown and plunges some 45m to form Taylorgill Force. In height and impact, this is perhaps the major Borrowdale fall.

To reach the fall, negotiate the arch in Seathwaite Farmyard (as for Sourmilk Gill), cross the bridge then turn left and follow the Derwent upstream by a rugged, but level track. Note the trout farm across the river. In less than 1k the track rises and bears right to enter the ravine where the slim column of the fall will be seen ahead. The force is formed by a bastion of stepped rock which spans the gorge to produce, in drier periods, a series of short falls that appear continuous down the centre of the crag and a small overflow on the extreme left. In very wet conditions the water projects outwards from the introductory cascades above to produce a continuous leap of nearly 30m. There is no plunge pool at its base but a gathering of angular boulders

Taylorgill Force, the highest single drop in Borrowdale

Sector One: Falls in Upper Derwent and Borrowdale.

hurled from above, an indication of its fury in spate.

Rowans and other trees clothe the left wall of the ravine and clasp the boulders in the rocky bed; high up on the left skyline a stand of conifers masks the main Stockley Bridge to Styhead path where passing walkers will hear the sound of the unseen fall below. On the right, or true left bank of the gill a steep, stony fenced track permits an alternative approach to Styhead.

Seathwaite to Taylorgill Force:- 1½k. (1 ml.)   Time:- 30 mins.

(11) SCALECLOSE FORCE
Borrowdale.   **Scaleclose Gill.**   **NY 245147.**

Parking is usually possible beside the road to Stonethwaite, off the Borrowdale Road. Wherever one chooses, the walk begins from the signed road to Longthwaite, opposite the Stonethwaite road end. Cross the bridge over the River Derwent and pass through the white metal gate of the Borrowdale Youth Hostel, there to turn immediately right, then right again to follow the track to Johnny Wood. Continue along the track, ignoring the stile entering the wood, to the second gate. Continue uphill across open grassland and bear left as the track angles towards Scaleclose Gill and a National Trust sign where the track descends to a footbridge, crosses the gill and heads towards Scaleclose Coppice which may be visited later. The fall however, is further upstream so take the ascending path from the NT sign until a convenient viewing stance is reached. See now the three stage fall of some 5½m. which ends in a small plunge pool before cascading noisily to join the Derwent.

The path continues uphill to a ladder stile and the Allerdale Ramble path. Left will take one to Seatoller or to join the Honister Pass. Right is the preferred direction; ½k will bring one to Tongue Gill but turn right and downhill just before the gill and follow the flow down to the Derwent. Cross the river by the New Bridge then follow it against the flow until the path can be taken to

Exploring Lakeland Waterfalls.

Rosthwaite. Take the path south from the centre of the village to Peat How to arrive at the starting point at Stonethwaite road end.
Stonethwaite road end and return:- 5k.   *Time:- 2 - 2½ hrs.*
There are many other delightful moderate or low-level walks in this area and an ascent of Castle Crag (NY 250159) is worthwhile.

(12) MOSS FORCE, NEWLANDS HAUSE

Borrowdale.	Moss Beck.	NY 194175.

This is the most accessible major fall in Lakeland, available equally to the elderly or disabled as to the young and hale, with the proviso that the former have private motor transport available. The fall lies at the summit of Newlands Hause, the pass between Buttermere Village and Borrowdale. There is ample parking by the fall; all may be viewed from the driving seat. The approach from Borrowdale via Braithwaite, Rigg Beck and Keskadale affords views of the fall ahead for the final kilometre, but on the short, steeper approach from Buttermere it is hidden until the summit is finally reached.

Moss Force viewed from the roadside.

Sector One: Falls in Upper Derwent and Borrowdale.

The feature consists of three separate series of cascades extending for something over 200m. from the skylined lip to the little ravine at its base. Depending on rainfall it may present a gentle, soothing swish of water over smooth, mossy rocks or link to form an outrage of leaping white water from top to bottom. When winter stills the flow, ice presents a challenge to climbers but beware the skim of ice that offers little anchorage.

The energetic may opt to return by way of Robinson, the Scope Beck, Low High Snab, Gillbrow and the road to Rigg Beck. The not so energetic may well retire to the Bridge Hotel, Buttermere, to discuss the next challenge in the bar.

(13) FALL ON SCOPE BECK
Borrowdale. **NY 216178.**

The valley of the Scope Beck and Littledale, the small comb at its head, lies between Robinson and Hindscarth and their respective ridges which are much more frequently visited than the remote vale below. There is an abrupt change of level between the upper and lower sections of the valley which causes the fall where all the tiny streams of Littledale amalgamate and shoulder through a rocky knick barely a metre wide, to plunge downwards to form Scope Beck.

A convenient starting point is from a small lay-by (234195) near Littletown where, south of the hamlet, the narrow road dips before curving right to Chapel Bridge (231194). Walk back up the road to the stile and cross the fence, then turn back on one's self to walk south with Knot End above on the left and the lay-by on the right, heading into the Newlands Valley. In about 1k the abandoned Goldscope mine workings will be seen as a scar on the fellside to the right across the Newlands Beck. A path branches off to cross the beck by a footbridge. Bear right to join the track to Low Snab Farm, to which there is no public access and bear left to follow the rising boundary wall. Avoid the path that strikes left up the ridge to Hindscarth and continue rounding to the left into the east flank of the valley of the Scope Beck. The path is thin but clear. Ahead, one will become aware of the V-shaped knick on the skyline where, in wet weather the head of the fall may be seen descending from Littledale. On the occasion of the writer's first visit to this strange dale, a wet and miserable day, the knick and its fall, amplified by mist, appeared suspended from the overlying clouds, giving the scene an aura of mystery as if belonging to another time and place.

On the valley floor is the level, grassy dam of a small reservoir with the white dash of its spillway, once used in conjunction with mining operations in the dale traces of which can be seen beside the path. The beck should be crossed at the dam to the west side to avoid the Littledale Crags which rise above the beck on the east. The crags sustain a variety of vegetation in their damp, dark crevices, from bracken to rowan and - would you believe it - dog rose and

The Scope Beck descends in a series of cascades from Littledale through the "V" shaped "knick".

Sector One: Falls in Upper Derwent and Borrowdale.

honeysuckle. Here on the west flank of the valley it is steep and grassy with rock outcrops but the mine track affords an excellent overview of the whole downfall as it descends from Littledale in a series of three or four stepped falls separated by extended cascades and a fine long water chute. One may continue southward through the upper valley on sheep trods to Littledale Edge, overlooking Honister Pass. From here, turn left for Hindscarth and Dalehead or right for Robinson, or return by the outward course.

From the lay-by near Littletown to the falls:- 3k.(2 mls.)   Time:- 1 hr.

(14) FALL ON NEWLANDS BECK

Borrowdale.   **Newlands Valley.**   **NY 230161.**

The Newlands Beck drops into a hole!

The Newlands Beck flows northwards from its main source in Dalehead Tarn, flanked by the ridges of Hindscarth, High Spy and Maiden Moor. The approach begins as for the falls on Scope Beck but instead of crossing the Newlands Beck continue upstream on the true right bank, keeping to the old mine track. Pass the Climbing Hut and 800m. further on, at Castlenook mine workings where the track divides, take the well-graded path that angles upwards on the eastern slope, well clear of the major fall ahead where a close approach in poor visibility could prove dangerous.

The fall occurs where a sudden valley step causes the beck to drop into a rock basin, elbow sharply to the west then resume its northward flow in the valley bottom, leaving in its wake a further five smaller falls and attendant pools, venues for summer bathing. The feature is best approached from downhill, in line with and just below the path from where one may safely look across the basin to the approaching beck, see its 18 - 20m drop and watch the flow sweep away to the right towards the bottom of the valley. The head of the basin is on the left hand where, on its glistening black wall in season, a mass of brilliant yellow lesser spearwort grows.

From here one may continue up the narrowing upper valley to Dalehead, first returning to the main uphill track. On the way, enjoy the many cascades, still pools and little ravines that mark the upper course of the Newlands Beck. This exploration could well be combined with that of the Scope Beck but if so, the downward leg should always be walked with care.

Lay-by near Littletown (234195) to Dalehead Tarn:- 4k.   Time:- 1 hr.

(15) FALLS ABOVE ELLERS BECK
Borrowdale.   **NY 245175.**

Parking is possible in a lay-by on the east side of the Borrowdale Road just north of Grange Bridge or alternatively ½ k. south at the Bowder Stone car park. Cross the bridge and walk through the village. Ahead, on the fellside the wooded cleft that conceals the fall may be seen. Bear right along the metalled road to where, almost opposite the Borrowdale Gates Hotel, there is a gate and a signed footpath. Pass through and follow the path which bears diagonally left up a rise and heads for a gap in the wall. Pass through, noting the waymark which indicates the path (not the rough road) skirting High Close.

Sector One: Falls in Upper Derwent and Borrowdale.

Bear left of the Sewage Works and follow the fence to the gate and onto the Cumbria Way footpath. Bear left then uphill, to follow the left bank tributary of the Ellers Beck upstream to the rocky cleft. At the bottom is a concrete reservoir with warnings of deep water; the fall itself lies far back in the cleft, concealed by trees.

Ascend the track beside the cleft; the beck feeding the fall cascades down the fellside before plunging in a series of four steps totalling some 22m to the final plunge pool and overspill cascade. Owing to the vegetation and slippery, convex grass slope, great care must be taken when attempting to view.

The path continues uphill to Maiden Moor.

Grange Bridge to fall:- 2k.   *Time:- 45 - 50 mins.*

(16) HIGH FORCE AND LOW FORCE
Borrowdale.   Coledale Beck.   NY 193214 and 197215.

The Coldale Valley ends abruptly in the massive cliff of Force Crag; above is a cirque-like basin with a steep back wall, the feature being a typical product of glaciation. Further back in geological history, the area was extensively mineralised. This latter factor is the reason for the mine buildings here, now abandoned. The mine has been worked on and off since the 19th century and probably before that, producing lead, zinc and latterly barytes. The steep walls of the cirque and valley step have resulted in the two falls High and Low Force, so named because of their position in relation to one another, not because of their magnitude.

To reach the falls, parking space near the bottom of Whinlatter Pass at Braithwaite is sometimes available at the beginning of the Force Crag mine road (227236). Walk along the nearly level track beside the Coledale Beck towards the old mine workings. On a December day of hazy sunshine after a week of rain the white streak of Low Force was prominent against the dark face of Force Crag beyond and above the abandoned, eyeless buildings. An untidy succession of falls and cascades tumbled down, across and

Exploring Lakeland Waterfalls.

down again as it negotiated something over 250m of descent. Viewed from the mine road beside the buildings, white water appeared over the lip of the cirque to descend the crag in a series of steps for a third of its total height, disappeared within a right slanting gully for almost another third and reappeared in a further series of steps and chutes to curve between spoil heaps and enter Coledale Beck.

High Force is reached by crossing the beck on large, square slippery stone slabs, formerly used as a ford for wheeled vehicles and consequently a wet wade in flooded conditions. Follow the rocky track upwards and right towards Coledale Hause. The cirque, the upper mine buildings and High Force will be seen away to the right; the fall itself cascades some 50m down a gully ending in a pretty apron fall. The water flows across the floor of the cirque as the Pudding Beck before its second and greater plunge. As the catchment is limited the Forces are normally modest in volume so that it was a exciting to see them in full flow.

To make a circular walk returning to Braithwaite via Grisedale Pike, follow the path up to the head of Coledale Hause then right or north to Hopegill Head from where the ridge path right above Hobcarton Crags leads to Grisedale summit. The descent from here, by the steep, rocky right hand ridge is tricky especially in mist, but the track is well worn, the start being due east initially.

Braithwaite to Low Force:- 3½k. (2mls.)   Time:- 45 mins.
Low Force to High Force:- 300m. (1000ft.) ascent.
High Force to Braithwaite via Grisedale Pike:- 5k and 290m. ascent.
Time:- 2½ hrs. (Note that Falcon Crag has been the site of several accidents).

(17) FALL ON WYTHOP BECK
Upper Derwent.   **Thornthwaite.**   **NY 213284.**

The A66 from Keswick towards Cockermouth divides into a dual carriageway at Beck Wythop alongside Bassenthwaite Lake. Just 1k

Sector One: Falls in Upper Derwent and Borrowdale.

from its start the back road from Braithwaite via Thornthwaite joins the A66 at Woodend Brow (218277) where there is a car park, a "Picnic Site" and bus link with Keswick.

From the car park walk northwards along the section of old road to Beck Wythop where there is a bus turning area and two houses. On the west side of the road opposite the houses there is a stone bridge parapet beneath which flows the Wythop Beck. An unsigned but well-marked path follows the left bank upstream for about 250m to a point where the fall may be seen ahead. The feature consists of 6 - 8m of cascades and water slides. It is not practical to proceed further than the water tank due to the steep banks, lightly clothed with conifers and tumbled tree trunks.

Woodend Brow to Fall:- 1½ k.   Time 30 mins.

There is pleasant walking in the surrounding forest north of the bridge but a map obtainable from the Whinlatter Visitor Centre and a compass would be advantageous.

(18) FALLS ON BECKSTONES GILL
Upper Derwent.   **Thornthwaite.**   **NY 214267.**

The main merit of this expedition is that it permits a closer look at the whitewashed Bishop Rock and the Clerk without having to climb up to them, the former being a prominent landmark on the steep hillside seen from the A66 above the Swan Hotel. It also provides access to the heights of Barf and Lords Seat (552m.) with fine views across Bassenthwaite. The gill itself does not provide falls of great merit but merely affording a bumpy ride for descending water as any other gill would. It is steep, partly fenced and difficult to access. The approach is from the convenient car park beside the Swan Hotel (221265) at Thornthwaite. Walk up the metalled lane opposite the Swan, to pass through the first gate on the right where a sign warns of falling rocks; this leads to the whitewashed Clerk. The Bishop can be seen far above on the right. Continue on the left bank of Beckstones Gill and cross above

Exploring Lakeland Waterfalls.

the 1m water slide to join the main way-marked path. One is now isolated from the gill and it is not encountered again until the forest is finally left when it is crossed on a plank bridge at a wire fence and stile.

The main fall is a short way back downhill. Retreat some 30m downhill and venture right when facing uphill, through the wood across a carpet of slippery brown needles towards the fence and gill; let your ears be your guide to the 2m. fall and long water chute.

Return to the bridge and stile, cross and curve right to the summit of Barf. Lords Seat lies 1k WSW from here by the same path.

Swan Hotel to Barf:- 1½k. and 300m. (1000ft.) ascent.   Time:- 50 mins.
Barf to Lords seat:- 1½k.   Time:- 20 mins.

For the full story behind the Bishop, consider visiting the Swan Hotel! *(What's this I hear: an hotel no longer?)*

(19) FALLS ON SLADES BECK

Upper Derwent.   **Millbeck.**   **NY 261268.**

Millbeck lies north of Keswick on the lower slopes of the Skiddaw Massif. From the large roundabout on the A66 take the A591 and turn immediately right on the minor road signed Applethwaite; continue through the hamlet towards Millbeck where a lay-by beside a footpath sign for the Allerdale Ramble affords possible parking, (254263). Walk back towards Millbeck Farm and take the signed footpath on the north side of the road. Pass through the latched gate and follow the well-marked path on the right bank of the Slades Beck. Branch onto a minor track which slants downwards to the right towards the beck, where there are two small falls or weirs of 1½m. in height which look most impressive in spate. From here, the beck flows out of sight downstream through a narrow defile beside Benny Crag.

Return to the main path and follow upstream, over a stile and across a small tributary to where the valley broadens out. Here are

Sector One: Falls in Upper Derwent and Borrowdale.

two dams or weirs of stone and concrete extending across the width of the beck, each producing an apron fall of 3-4m. If their original function was to contain water, this can no longer be their purpose, for the containment area above each dam is entirely filled with loose stone scree. Their only use now is to deter erosion of the beck.

From above the upper dam, the valley narrows and becomes steeper and is filled with a succession of noisy and energetic cascades. Tracks on either side may be followed upwards, that on the left bank leads to a small railed concrete reservoir from which a pipe extends downhill, probably to maintain the service reservoir above Millbeck; here too is another small fall. The right bank track may be followed to Black Beck and Tongues Beck, their combined waters forming the bulk of Slades Beck. The main Gill becomes virtually dry except in very wet weather and could, as a penance, be followed steeply uphill to Carlside Tarn; but there are more pleasant routes.

Millbeck to upper dam and fall:- 1k.   *Time:- 45 mins.*

(20) FALLS ON KILNHOW BECK AND GATE GILL
Upper Derwent.   Threlkeld.   NY 317258 & 325263.

A visit to these two falls may be combined to offer a pleasant half-day circular walk to the north and east of Threlkeld. The village also offers the benefit of two excellent hostelries. Drive or bus into the village which lies just off the A66 from either the Penrith or the Keswick direction. From the centre, a steep road leads north towards the Blencathra Centre where 400m uphill there is a small convenient car park on the right.

Take the signed footpath alongside the Kilnhow Beck and cross by the timber walkway and bridge. The small 2m fall can be seen about 20m upstream from the bridge. Once across, continue along the footpath northwards to the gateway and take the signed path for 1k right, or eastwards to Gate Gill via a stile. Here a weir

Exploring Lakeland Waterfalls.

creates an artificial fall and affords a crossing of the Gate Gill to the start of the Halls Fell Ridge path. Visible 30m upstream on the right bank are the remains of the old lead mine buildings. Above these, the beck has been constricted by walling the right bank to provide a track upstream to the old levels. Here the water cascades impressively in spate.

However, the main fall on the Gate Gill occurs in a gloomy ravine partially obscured by trees, its base polluted with rubbish. To find it, follow the gill downstream from the weir and pass through the latched gate to where the stream disappears into the ravine on the left. Beyond the wall and fence it reappears as a series cascades falling in total some 15m.

After enjoying this somewhat noisome spectacle continue southwards to the minor road which heads right back to the village. Bus stops are handy or walk through the village, past the two hostelries and turn right up the hill to regain the car.

From Threlkeld to falls and return:- 3k. (2 mls.)   Time:- 1½ - 2 hrs.

<><><><><><>

Sector Two:

Falls Back o'Skiddaw.

This sector is separated from the main "dome" of Lakeland and its radial valleys by the east to west vale ploughed by an ancient glacier heading perhaps towards what is now the Irish Sea, but presently occupied by the Rivers Glenderaterra and Greta and conveniently used by the A66 Trunk Road. The latter serves as a useful boundary to this Sector.

The rock here is mainly Skiddaw Slate with underlying granite which is exposed in places by weathering (in the bed of the River Caldew for example) once yielding deposits of lead and copper and some more exotic minerals such as tungsten, all now virtually exhausted, leaving a legacy of old mine buildings, tips, shafts and adits.

An important geographical feature encountered in this sector is the great fault, a north to south depression that separates the Skiddaw Massive from Blencathera and extends all the way to Morecombe Bay by way of St. Johns in the Vale, Thirlmere, Grasmere, Rydal and Windermere. Great Calva stands at the head of this glacially modified trough. The main falls in this sector are:-

Fall no.
- (1) Whitewater Dash.
- (2) Southerndale - falls on.
- (3) Roughten Gill - falls on.
- (4) Roughton Gill - falls on.
- (5) Brandy Gill - falls on.
- (6) Charleton Gill - fall on.
- (7) Glenderamacken fall on.
- (8) The Howk, Caldbeck.

<><><><><><>

Exploring Lakeland Waterfalls.

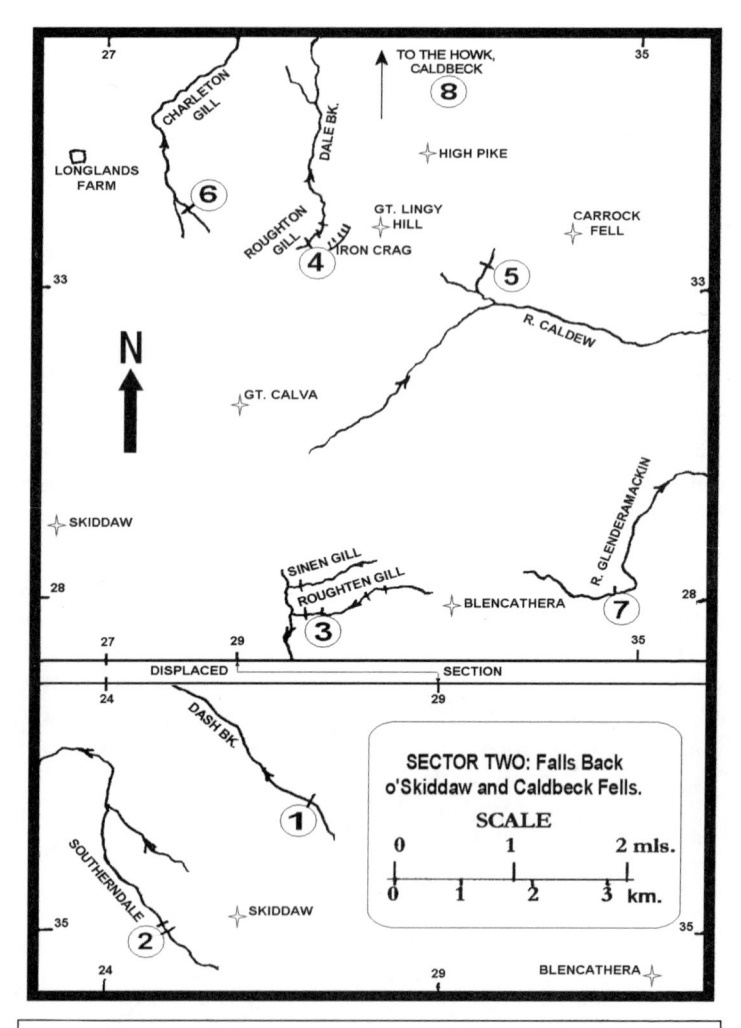

Sketch map to show the location of falls (1) to (8) in Sector 2.

Sector Two: Falls Back o' Skiddaw.

(1) WHITEWATER DASH

Back o' Skiddaw.   **Dash Beck.**   **NY 273314.**

At 80-85m in height Whitewater Dash in full flow is spectacular, but as this is spread over a horizontal distance of about 120m, in three series of cascades, it is therefore not a vertical, unsupported fall as it may appear when viewed face on from the approach road. It is in fact, well laid back at an average angle of about 38 degrees, a long series of cascades, chutes and falls, these components merging to form a wonderful spectacle which the eye may view in its entirety as it descends the great step at the head of the valley. Except, that is, for the top section around the corner; which is also a fearsome force in its own right, launched from a rock platform to foam down blue-black rocks for 10-12m.

The route to Dash Beck is by the Orthwaite Road; turn off the A591 just before reaching Bassenthwaite when travelling from Keswick. Park in a lay-by beside Peterhouse Farm (249324); from here a gated road, part of the Cumbria Way, leads across fields to the head of the falls and on to Skiddaw House. Before climbing the steep upward path, take time to follow a slim trod track to the pool at the base of the falls to sense their power and thunder.

From the top of the fall a fence extends SW and uphill to Skiddaw via Bakestall, a far more rewarding ascent than the tourist

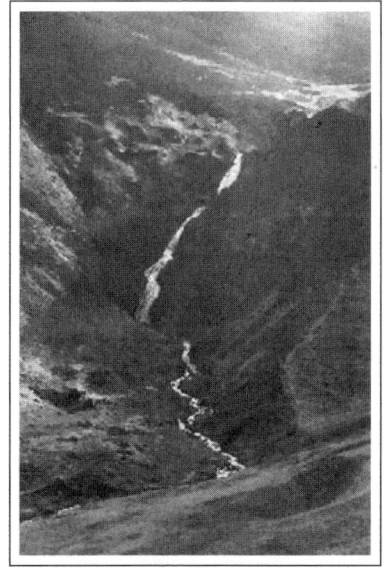

Dash Beck descends in a series of powerful cascades.

Exploring Lakeland Waterfalls..

route from Keswick via Latrigg.
From Peterhouse to the fall:- 2k.   *Time:- 30 - 40 mins.*

(2) FALLS ON SOUTHERNDALE BECK
Back o' Skiddaw. **NY 245298.**

Few walkers visit Southerndale, there being little to attract them, blocked as it is by the daunting bulk of Longside Edge and Skiddaw. Those who strike up the Edge towards Ullock Pike may afford it the occasional glance downwards to the left, but to those who favour seclusion, accessible but remote, this is the very place. It has an objective too, in its two small waterfall combinations; an ideal venue for a summer half-day stroll, a laze in the sun with the music of falling water to lull one to drowsy relaxation.

From Keswick take the A591 towards Bassenthwaite Village and at High Side House turn off right onto the signed Orthwaite Road. Park in a large but popular lay-by about ½k after the junction. Beyond the lay-by a gate right, gives access to a path which angles left towards Barkbeth then is way-marked right beside a line of spaced thorn trees that edge a green road which swings left again towards a ladder stile. Note the mound of Barkbeth Hill surmounted by a seat. Continue to and cross the next ladder stile. Follow the green path as it curves gently to the right and becomes a gravel track as one approaches a stile where the verge drops steeply to Southerndale Beck. Cross the plank footbridge and continue along the green road that follows the beck into the dale. Before crossing the footbridge note, for future reference, the easy path that angles upwards towards a col on The Edge just beyond the Watches; an easy approach to Ullock Pike.

After a kilometre of easy walking the track nears the beck and the sound of rushing water indicates the presence of the falls. There are two combinations some 30m apart, the higher being the more vigorous cascades with falls of 2m. Above the cascades the blunt ridge of Buzzard Knott on the Skiddaw side separates

Southerndale from Barkbethdale; on its flanks the remains of exploratory mine workings suggest a reason for the existence of the green road.   *Distance:- 2½k.*   *Time:- 50 mins.*

(3) FALLS ON ROUGHTEN GILL
Back o' Skiddaw.   **NY 305278.**

The Glenderaterra Beck flows south between Skiddaw and Blencathra, occupying the great fault that separates the two fells until it is trapped by the deeper valley of the R. Greta midway between Keswick and Threlkeld where its waters are swept west to the Irish Sea.

A broad track loops into the fault from the Blencathra Centre where there is a small car park (302257) which may be reached from Threlkeld; walk along the track to where Roughten Gill flows from the slopes of Blencathra, beneath a slab bridge and so into the Glenderaterra. On the way, note the small, stepped fall of about 4m (at 299269). Roughten Gill itself occupies a narrow groove which may have been associated with the original faulting causing weaknesses that developed into the hollows and clefts where four sets of falls now occur. The hollows offer sheltered environments where trees and other vegetation thrive on this otherwise barren fellside. It is the two upper falls which are of main interest, the lower with rocky cascades, the higher a slit with narrow, stepped falls of 3-4m.

From the top fall the choice is to continue eastwards to the summit of Blencathra and return by the main path SW above Knowe Crags or follow the level contours in a southerly direction to join the same path lower down and regain the Centre car park.
Blencathra Centre to top falls:- 3½k (300m. ht.)   Time:-1½ hrs.

Sinen Gill, 250m northwards from Roughten Gill also yields a small fall that may be seen from the main track.

Exploring Lakeland Waterfalls.

(4) FALLS ON ROUGHTON GILL
Back o' Skiddaw.   **NY 303344.**

Lest there be confusion, it should be pointed out that Roughton Gill referred to here is distinct from Roughten Gill (previous page) that drains the western slopes of Blencathra and flows into the Glenderaterra Beck.

Our Roughton Gill is a tributary of Dale Beck, accessible from Fell Side (305376) where parking is possible, but reached from Caldbeck or Hesket Newmarket. Just beyond Fell Side, the turning by Fell Side Farm leads to an area of wide verges where parking is possible.

Head for the hills along the remaining section of metalled road to a stile and follow the signed rough road northwards to cross the Ingray Beck by the footbridge. The area is rich with the remains of former mining for lead and copper and spoil heaps still contain mineral samples, though much ravaged by collectors. In order to preserve the remaining samples collectors are now required to be licensed .

The track approaches Dale Beck and crosses by a footbridge to the right bank. The valley narrows and progress becomes difficult where erosion of the steep flanks has caused large sections of turf to slip exposing precarious shale and slippery rock to provide unsure footing. A succession of steep tributary gills flow in: Hay, Ramps, Birk, Clints and Silver, each one containing abandoned mines or exploratory levels.

At the confluence of Silver and Clints Gills the valley widens; here are the mounds of more spoil heaps. Ahead is Roughton Gill emerging from what has narrowed to become a virtual gorge. Follow the narrow path on the right, or true left bank and look down with care to see the first fall, a slender ribbon of white water with a drop of 4-5m. Continue upwards beside the cascading beck, noting the mine level that gapes darkly on the right of the flow. There follows an almost vertical wall of rock down which the

Sector Two: Falls Back o' Skiddaw.

stream descends with a satisfying "shush". Far above is Iron Crag. Pass the fall by the grassy slope on the right-hand side to negotiate the final elongated cascade, which pauses briefly before continuing the descent.

Having extricated oneself from the gill and in preference to retracing one's steps back down the awkward gorge, head east around Little Lingy Hill and north-east on the Cumbria Way towards Great Lingy Hill where on the SE slopes one may seek temporary shelter at the old shooting box at 312336. Continue over High Pike and down its northern slopes towards Fell Side - being careful to avoid the old mine shafts.

Fell Side to head of RoughtonGill:- 4k.   *Time:- 1½ - 2 hrs.*
Return distance via High Pike:- 5k.   *Time:-2½ hrs.*

(5) FALLS ON BRANDY GILL
Back o' Skiddaw.   **NY 323335.**

Halfway between Threlkeld and Penruddock on the A66, a road signed "Mungrisedale" extends north and skirts the eastern flanks of Blencathra. Beyond Mungrisedale, with its hospitable Inn and St. Kentigern's Church, is Mosedale. Here the River Caldew flows out of its wide valley and swings north on its journey towards Carlisle.

Towering above Mosedale is the granite bulk of Carrock Fell, topped by an iron-age fort. Turn sharply west or left here and head up the valley, past Swineside with its dwelling and knot of trees to the confluence with Grainsgill Beck; here it is possible to find a space for parking.

The path beside the infant Caldew leads to Skiddaw House and eventually via the Glenderaterra valley to Threlkeld or Keswick, a delightful walk. The way to Brandy Gill however, is to follow Grainsgill Beck upstream to the remains of the mine buildings; just beyond, the target stream flows in from the right, or north. Take care to avoid open shafts while heading upstream. The instability of the gill sides means that the stream has to be crossed from side to

Exploring Lakeland Waterfalls.

Brandy Gill falls from beneath a natural bridge formed by a collapsed rowan.

side to avoid the minor scree falls. In normal conditions the flow is very modest, only a stride wide. Indeed, the falls when reached are modest indeed, the first a slim but pretty stepped cascade of 5-6m in total and easily recognisable by the collapsed rowan that lies across its lip - a distinctive feature. The upper fall is smaller still, a mere 3m but beside it is a rectangular mine entrance; within, an echoing passage deep in water.

Grainsgill Beck to top fall:- 1k.   *Time:- 35 mins.*

(6) WATERFALL ON CHARLETON GILL
Back o' Skiddaw.   **NY 282352.**

Seen some weeks previously from a distance, magnified by mist and the dusk of a winter's afternoon, the white ribbon of water gave the impression of a considerable fall. Time and a rendezvous prevented immediate exploration so it was disappointing in the later proving visit to discover that the drop was a mere 2 ½m., the flow sparse and fragmented as it descended a steeply stepped cascade. The Charleton Gill with its deeply entrenched watercourse, shares this characteristic with many of the streams in this area which suggests a more vigorous period in their post-glacial past.

To reach the fall, park by the water-treatment works near Chapelhouse Reservoir below Longlands Farm (266358) and follow the Cumbria Way beyond the gate to pick up the old bridle path that heads south and upstream beside The Charleton Gill. At the confluence of the two minor gills that become the main gill, the fall will be seen a short distance upstream on the right bank tributary.

Longlands lies on the Orthwaite Road which branches north off the A66 at Side House between Keswick and Bassenthwaite and 1k north-east of Overwater.

Longlands Farm to the fall:- 2½k.   *Time:- 45 mins.*

(7) FALL ON THE GLENDERAMACKIN
Back o' Skiddaw.   North of Mousethwaite Col.   NY 347281.

The wandering Glenderamackin draws its waters partly from Scales Tarn, held in the comb below the notorious Sharp Edge, east of Blencathra summit and the neighbouring slopes between The Edge and Bannerdale Crags. The water flows south-east then turns sharply in a northerly direction for a couple of miles to circle Souther Fell at Mungrisedale and return southwards again. The fall itself is located at the first sharp bend at the start of the river's northerly leg.

The shortest approach is to park at a lay-by on the A66 at Scales where the White Horse Inn is situated and take the back road east from the Inn to Mousethwaite Comb. Here the minor road is gated and there is parking for two or three vehicles only. This road continues to Mungrisedale and is a possible return route to bear in mind.

Take the graded path into the comb, rise steadily and slant right

Exploring Lakeland Waterfalls.

to the col. Below is the river; descend leftwards on the path to the footbridge that crosses it and note the 2m fall just downstream that drops into a curved rock basin. The main fall is some 200m further downstream where rock outcrops and overhanging trees mark the fall; the roar of tumbling water is distinct as one approaches. Descend from the track with care in wet, slippery conditions as the area is dangerous.

The fall is not easy to observe from the grassy platform above for the water is partially concealed within a steep chute which is angled away from the viewer. At the bottom of the 10m chute is a circular rock basin overhung by a prominent rowan, which spreads its branches as if to claim the rights to this fine bathing pool.

From here one may continue along the path beside the river at the base of Souther Fell to Mungrisedale and its hospitable Inn and return by the minor road to the car or explore upstream to Scales Tarn. Other more adventurous routes may come to mind.

Car park on the A66 to main fall:- 2k.   *Time:- 1 hr.*
Fall to Mungrisedale:- 3k.   *Time:- 1 hr.*
Return via minor road flanking Souther Fell:- 4k.   *Time:- 1½hrs.*

(8) THE HOWK: CALDBECK
Back o' Skiddaw.   **Parkend Beck.**   **NY 320397.**

On the very northern edge of the Lake District National Park is the village of Caldbeck, once a centre of industry based on the mineral wealth from the nearby fells to the south; associated industries using water power such as a woollen mill, a bobbin mill, a saw mill and notably a brewery and corn mills to supply the essentials of life. During the 18th and 19th centuries the little town thrived, but come the 20th and the exhaustion of the mines it lapsed into the peaceful village of today. There are many attractions for the visitor: gift and craft shops, one remaining Inn of the dozen or so that once sustained the miners, beckside walks, the lovely St. Kentigern's Church with its Norman doorway and

Sector Two: Falls Back o' Skiddaw.

nearby, the famous huntsman's grave.

The village also provides a quiet day for fall hunters away from the energetic demands of the fells, for it is sited on the rim of carboniferous limestone that encircles much of the Lake District. The Parkend Beck flows from west to east through the village, in doing so it has eroded the softer shales downstream (to be seen in the beck beside the Church) leaving proud the more resistant rock upstream where a typical limestone gorge has been carved. The head-ward erosion back through the gorge has resulted in a step and the fall so formed once supplied power for the old bobbin mill. To visit The Howk as the area is known, is a must. From the small central car park leave by the exit towards the Duck Pond; turn left to the junction, cross the road and follow the signed path. First one sees the well-preserved wood drying shed, then the bobbin mill itself where some restoration work has been done. Once a great iron waterwheel supplied power for the lathes but this went as wartime scrap in 1940. The gorge narrows to where a walkway, steps and footbridge give views to the watercourse below. Here are alleged to be two holes known as the Fairy Kettle and the Fairy Kirk down which the water in spate tumbles and churns. Observation in normal weather showed the water sluicing through narrow horizontal and vertical channels to produce a fall of 2 - 3m..

Car park to the top of the Howk:- under 1k.

<><><><><><>

Sector Three:

Falls in the Thirlmere Valley
(Eastern Flanks).

The Thirlmere Valley extends northwards from Dunmail Raise and continues as the Vale of St. John's to Threlkeld, all part of the great fault mentioned in the Back o'Skiddaw sector, which extends from Great Calva in the north to the tail end of Windermere. The flanks of this ancient fault, albiet much modified by Ice Age glaciers, is virtually indistinguishable from any other glaciated Lakeland valley, and makes this an ideal site for waterfalls with its steep fellsides where one will find a concentration of falls of good quality.

The reservoir itself was constructed in late Victorian times with the foresight that increasing demands for water in the Manchester region would need to be addressed. The building of the dam, completed in 1885, drowned two small existing lakes, some farms and cottages; only Wythburn Church survived. Halfway along the lake beside the A591, is a castellated, hexagonal building with its straining well sunk deep below the lake surface where the water is "strained" through filters before entering the pipeline to be directed by gravity southwards towards the Manchester conurbation. Surplus water from the reservoir, released through the spillway from the dam at the northern end, flows out as St. John's Beck to join the Glenderamackin at Threlkeld then as the River Greta, flow through Keswick to join the outflow from Derwentwater, already the River Derwent and eventually to the Solway Firth at Workington.

The A591 road from Grasmere to Keswick follows the eastern side of the reservoir. A minor "tourist" road skirts the western shore where there are several small car parks and access to the lake path; the road is signed at the bottom of Dunmail Raise and across the dam at the north end. The fellsides on either side of the valley

Sector Three: Falls in the Thirlmere Valley.

are clothed with conifer plantations where a small population of the rare red squirrel survive. As these pretty creatures lack road sense a number of ropeways have been slung across the A 591 in an attempt to reduce fatalities.

A permissive forest track extends parallel with the main road from Birkside Gill, through the band of forest to the car park and nature trails at the Swirls, crossing on the way all the gills that flow directly into the reservoir. A path also links the Raise Beck with Birkside Gill. Yet another links The Swirls, across the bare fellside northwards above the King's Head, Thirlspot, with Stanah Gill where the water from each stream in this section is scooped by a concrete channel and relayed to the reservoir at the Swirls. The falls in St. John's Vale and the Mosedale Beck on the contrary slope which flows north to the Glenderamacken, are also included here.

There are four points of access to the forest track: (1) through the Swirls car park, (2) from the Wythburn Church car park, (3) via a gate at the bottom of the Raise and (4) along the path from the Raise Beck.

The main streams with falls on the east flank are:

(1) Raise Beck,
(2) Birkside Gill,
(3) Whelpside Gill,
(4) Miners Gill,
(5) Helvellyn Gill,
(6) Stanah Gill,
(7) Fisherplace Gill,
(8) Mill Gill and adjacent streams,
(9) Mosedale Beck.

Exploring Lakeland Waterfalls.

The OS map to cover Sector Three is mainly Outdoor Leisure 4 plus the western edge of sheet 5.

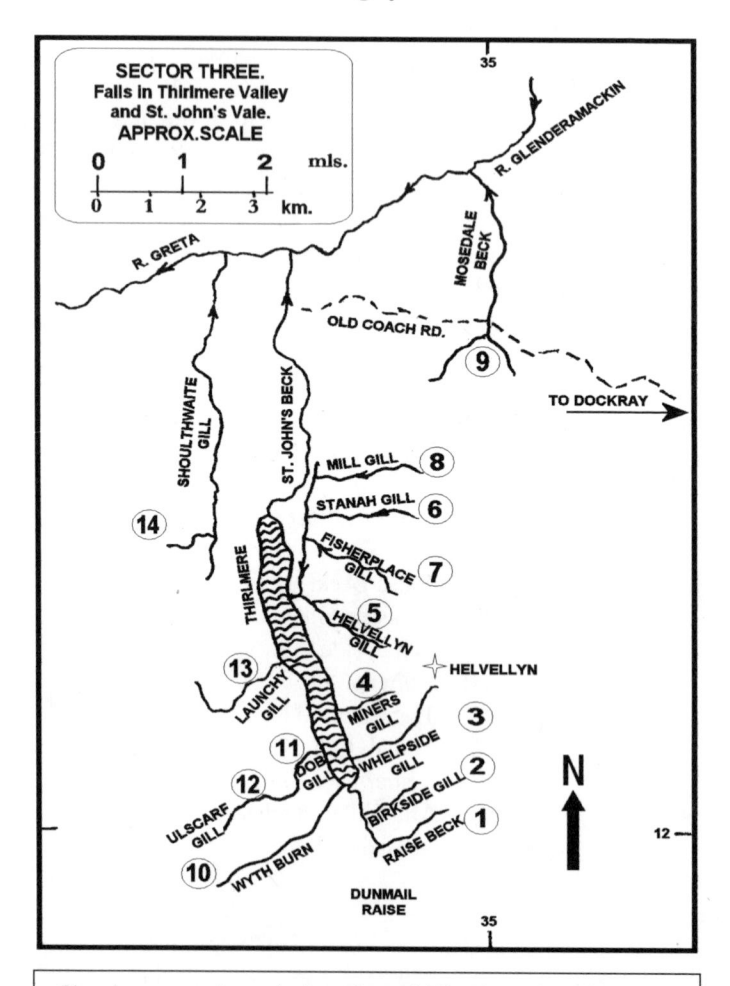

Sketch map to show the location of falls (1) to (9) in Sector 3.

Sector Three: Falls in the Thirlmere Valley.

(1) FALLS ON RAISE BECK

Thirlmere.   **East Flank.**   **NY335119.**

This temperamental stream descends from the fellside east of Dunmail Summit, gathering its water from Willie Wife Moor below Dollywaggon Pike and Seat Sandal. Just before reaching the road the beck forks, the main flow swings north towards Thirlmere, gathering the flow from Birkside Gill on the way. A trickle swings south, passes by a culvert beneath the A591 and heads south towards Grasmere, gaining strength from the gills on either side; there are consequently two Raise Becks flowing in opposite directions. The diversion northwards was a deliberate decision by the Manchester Corporation Water Authority in order to transfer as much water as possible into the reservoir. It was mainly the inability of the south-flowing branch to cope with the excess water and rock debris that blocked the culvert and the pass itself for several days, after a storm in the early 1990's.

    The Raise Beck is a popular ascent for walkers heading for Grisedale Tarn and the surrounding fells; icy in winter, severely eroded and awkward in descent at any time. Its two falls are modest, lively only on those rare occasions when it would not be a good time to be around in this restricted gully. The falls are well separated on either side of the midpoint; the lower a ragged 4m slide down a V-shaped groove eroded in a cross band of harder rock. Above is a 10m cascade where white water hurtles down a series of black rock steps.

    There is parking near the summit of the Raise on the Grasmere side just after the short section of dual carriageway and access to the beck from the road via three ladder stiles. A path can be followed to Birkside Gill and on towards the forestry track to the Swirls which links the other eastern streams.

Ascent of Raise Gill to Grisedale Tarn:- 45 mins.   *Height:- 340 m*

Exploring Lakeland Waterfalls.

(2) FALLS ON BIRKSIDE GILL

Thirlmere.   **East Flank.**   **NY 328125.**

This stream which gathers its water from Birk Side, below Dollywaggon Pike, often puts on a spectacular show, visible to traffic as it crosses Dunmail Raise. As it is just clear of the woodland, its wet weather performance of white cascades is the most easily visible of all the Thirlmere destined streams, but the narrow, walled road over the Raise permits no dallying; only the passengers can afford to enjoy a few moments of pleasure. It is the walker who may fully savour this sporting stream.

There are two main falls, the upper of about 3m, the other of 6m, with plenty of smaller cascades above, between and below, which appear to coalesce in one long foaming white display in spate, but are usually reduced to separate entities in average weather when the two main falls are reduced to narrow ribbons of water.

Distance from Dummail summit:- 800m   *Time:- 30 mins.*

(3) COMB GILL AND WHELPSIDE GILL

Thirlmere.   **East Flank.**   **NY 332140.**

From Birkside Gill the level forest road provides easy walking northwards, parallel to the A591 where traffic is faintly heard but unseen below. After 1k the Helvellyn track from Wythburn Church, access point (2) is crossed. The two streams cross the forest road some 200m further on, but the approach to view them is by way of the Helvellyn track which zigzags steeply uphill from the Church. Once clear of the trees continue alongside the forest wall until it diverges to the left. Here, contour left to cross Comb Gill, which has little of interest to offer, and around Middle Tongue to Whelpside Gill. This is a steep, bouldery, bracken-cloaked, ankle twisting approach but the reward after rain is a real beauty of a four-stage staircase, each fall of about 3m and

Sector Three: Falls in the Thirlmere Valley.

linked by rocky cascades with a finale of a long, swooping water-slide. From here, one may regain the Helvellyn path and return to the forest track for the next stream encounter.
Distance from Wythburn Church:- 800m   Time:- 30 mins.

(4) MINERS GILL
Thirlmere.   **East Flank.**   **NY 326148.**

From the point where the Wythburn - Helvellyn path crossed the forest road continue northward, cross in 200m the Whelpside Gill and in another 1k cross by separate footbridges, two more descending waters, the second and larger being Miners Gill (unnamed on some maps) with its source from below Helvellyn. Beyond the bridge a clearing permits fine views across Thirlmere towards the western flanks. Upstream, the water descends through a stock fence as a cascade, then a 3m fall, followed by further cascades until it rushes beneath the footbridge and below one's feet to continue its cheerful descent to Thirlmere. There is little of interest above the stock fence where there is a padlocked gate beside a rickety stile. Upstream there are some further cascades, an overgrown mine track which crosses a dry gill to mine tips and the remains of a stone building with four compartments. (All mining in the valley ceased with the formation of the reservoir.)
Distance from Wythburn Church:- 1½k   Time:- 30 mins.

(5) HELVELLYN GILL
Thirlmere.   **East Flank.**   **NY 323167.**

From the footbridge over Miners Gill, 2½k. of easy walking north along the forest track, will bring one to The Swirls (access 1) car park, public conveniences and picnic area. Once clear of the woodland, cross the car park to the signed path to Helvellyn. Follow this beside Helvellyn Gill for 200m to where a footbridge crosses it; a small fall is visible to the left where a tributary stream

Exploring Lakeland Waterfalls.

emerges from a tree-shrouded gill. The real falls on this little tributary lie some 60m upstream and are completely obscured and inaccessible from this point. To view, return across the footbridge and climb the fellside to a small group of boulders. Curve right towards the upstream section of the gill and descend into it with caution to view two near vertical falls, each of about 3½m in height and 1½m in width, overhung by larch and framed by ash. These falls would appear to be seldom visited as no trace of a path was visible. Other falls on the main Helvellyn Gill are small compared with these.

If one has not arrived at the Swirls by walking the forest road, the car park here is the obvious starting point for this fall. The time from here is about 30 mins.

<><><><><><><>

This concludes the first section of the Eastern streams; there is a considerable gap before the two final streams which will be described from a different starting point. This begins 2k north along the A591 towards Keswick from the Swirls car park, where the St. John's in the Vale road, the B5322, branches off right at Stanah. Here there is a lay-by on the main road opposite a bus shelter. Half-a-kilometre beyond the junction there is also a large Forestry car park at Legburthwaite (NY318196) on the St John's road.

(6) STANAH GILL

| **Thirlmere.** | **East Flank.** | **NY 320189.** |

This stream drains the western slopes of Watsons Dodd and Stybarrow Dodd towards the hamlet of Stanah and is captured in the concrete conduit or leat which carries it and the water of other gills in this section, for transmission to the reservoir.

From the telephone box at 318189 walk up the metalled road to its left turn; straight ahead is a ladder stile which is the start of Sticks Pass. This pass was once marked literally with a series of

Sector Three: Falls in the Thirlmere Valley.

sticks which guided miners across the fell to the Glenridding mines. Follow the uphill track, cross the leat and then the Stanah Gill by a wooden footbridge where, just downstream and partially concealed by trees is a 3m fall. At the fingerpost beyond the bridge keep left and uphill; the lower path leads towards Fisherplace Gill which will be visited later.

Twin falls in upper Stanah Gill.

Exploring Lakeland Waterfalls.

Continue uphill on the Sticks Pass track for about 200m avoiding the dangers of the little ravine on the left, until it is possible to edge through the bracken towards the watery delights of the gill. There is nothing of great magnitude within this water-garden, but rocks, cascades and waterslides of great variety may give pleasure - a gill scramble within the capability of the average booted walker, provided the beck is not in spate. Above these initial delights a 3½m apron fall may make progress difficult where the banks narrow and becomes steeper. Around the corner another stretch of similar cascades culminates in an unusual twin water feature comprising a domed, moss-green rock buttress from the crown of which a ribbon of water falls some 4m to burst into spray on a ledge; 90 degrees to the right a similar slim cascade spills over from the same source. This obstacle in the bed of the gill may be bypassed by scrambling up the scree to the left of the feature. The final section of interest consists of a staircase of five regularly-spaced cascades of 2 - 3m; beyond, there is little more of interest unless one wishes to face the hazzards of Brund Gill.

The main gill, visited when the afternoon sun strikes into its depths, is full of sound and sparkling beauty. If wandering within the stream-bed is found to be tedious most of its features can be admired from above by deviating at intervals a few metres from the Sticks path, which also offers a suitable means of return.

Distance:- 1k   *Time:- 1 hr.*

(7) FISHERPLACE GILL

Thirlmere.	East Flank.	NY 323182.

When exploring Stanah Gill one turned left and uphill at the fingerpost. To find Fisherplace Gill continue along the permissive path beside the fell wall, parallel with the A591 far below, for some 700m to a pleasant hollow of short, green turf where a footbridge crosses the gill, a magical place on a bright September day - and on most other occasions. From here falls and cascades could be traced

Sector Three: Falls in the Thirlmere Valley.

to the very skyline; the cool air vibrated with the rush and roar of water below the bridge where red rowan berries nodded to the shining gill. Far upstream, a slender column of white materialised, after a stiff uphill scramble, into an 8-9m fall received by a circular plunge pool, followed by another 4m drop below. To complete the display, the skyline fall further up the fellside is split by a huge projecting rock before tumbling 6-7m into an undercut hollow. Fisherplace Gill must be accorded merit as one of the finest and most picturesque water-displays in Northern Lakeland.

A continuation along the path beside the intake wall will carry one across several minor streams, some of which may become quite lively in very wet weather, until the wall and path turn downhill towards The Swirls. To complete the round cross the A591 at The Swirls to the large lay-by across the road noting the discharge into Thirlmere of the contents of the concrete channel. Now descend to the lakeside path then follow it northwards beside the lake to Greathow Wood. Skirt its eastern edge on the permissive path to the short lane that links back to the A591 near to where one is parked.

Stanah may be reached by bus from Keswick and return. Refer to local bus timetables at the Booths Supermarket, Keswick.

Distance for the round:- 7½k including falls.   Time:- allow 4½ hrs.

(8) MILL GILL

St. John's in the Vale.   Eastern Flanks.   NY 321198.

From the car park at Legburthwaite (NY 318196) gaze up across the road to the towering Castle Crag; to its left the deep scar of Mill Gill cuts the flank of The Vale of St John's as do several other streams along the length of the Vale.

Mill Gill is approached from the car park by following the path opposite the toilet block to the road; cross to the wooden access gate opposite and walk up the fell choosing from the several tracks heading to the left of Castle Rock, frequently the haunt of rock climbers. Cross the leat, or concrete conduit which carries the water to Thirlmere. Continue on the upward track to the wall and gap which provides access to the open fell. Follow the plantation wall round until Mill Gill is reached; cross to the true right bank, noting the weir. Above is a sheer-sided, narrow ravine where trees and shrubs cling to the mossy wall and noisy water emerges. Far back in the cleft is the largest fall on the gill, of about 5m, barely seen when leaves abound; below it several minor falls and cascades emerge from the confining cleft.

The gill may be followed upstream by a steep, little-used track well away from the convex slopes that fall into the ravine, which gradually widens to expose a double water chute. Beyond the chute a left bank tributary flows in at a point where the gill executes a double right-angle bend after emerging from another but less confining gorge containing cascades within. Further upstream still the beck flows within a V-shaped valley with evenly spaced series of small falls and cascades, the limit of interest in this gill as regards water features. From here one may wander on upward to Millgill Head and Great Dodd summit or contour northwards towards Calfhow Pike and Clough Head.

Legburthwaite to top of Millgill Head:- 2k   *Time:- 1 - 1½ hrs.*

Along the flanks of St John's Vale other gills descend; they are steep and dangerous with broken, rocky sides where an attempt to descend from above could be a serious health risk. There is really no public access across the farmland below. Owing to the lack of catchment the falls recorded on the OS map are intermittent and insignificant except in torrential rainfall. These gills are as follows;
1) Ladknott Gill; amusing and negotiable from Mill Gill. Water enters the head of the gill via a sub-surface stream as a 3m cascade and displays in miniature all the features practised by its larger sister in noisy precocity before it too, is lost to the concrete conduit.
2) Two unnamed gills descend to the rear of Fornside Farm - no

Sector Three: Falls in the Thirlmere Valley.

falls, except in very wet weather when it appeared water might well enter by the back door and emerge at the front.

3) Beckthorns Gill; cascades produced in heavy rainfall, but short lived.

4) Sandbed Gill; accessible from the level of Jim's Fold, where there are three thin falls of 2-3m, seldom visited but charmingly reminiscent of an artificial garden stream; a warm, sheltered spot on a sunny day, even in winter at 400m altitude.

Distance along the flank from Megs Gill to Jim's Fold:- 2½k

(9) MART CRAG AND ROWANTREE FALLS
Threlkeld Common.   Mosedale Beck.   NY 347225 and 345222.

Rowantree Beck and Rowantree Gill drain the northern slopes of Great Dodd and Clough Head and combine to form the Mosedale Beck, which flows northwards to cross the Old Coach Road, eventually to join the wandering Glenderamackin. The Coach Road extends east to west between Dockray and the Vale of St John's to cross the Mosedale Beck at Mariel Bridge at the midpoint of the road. Upstream from the bridge there are two sets of falls of interest. Mart Crag Fall is five minutes along the true left bank and consists of an apron fall generated by a 4m rock step; here the stream showers onto broken rocks then into a tree-shaded pool before resuming its turbulent progress. A further 20m upstream is a 3m, two stage cascade which elbows between water-worn, moss-blackened rocks.

From here the hint of a narrow but shallow ravine, edged by trees may be seen further upstream: Rowantree Fall should not be missed. Stand in the gravel stream-bed at the entrance to the sheer-sided rock defile where the water glides smoothly out and look to where the white curtain of the top fall thunders down, partly obscured by the nearer cascade which descends a water slide in the foreground. From this viewpoint the top fall gives the illusion of far exceeding its mere 2m. Above and beyond this upper fall the rock knoll and cairn of Calfhow Pike can be seen.

The traverse of the Old Coach road can form part of a challenging mountain bike ride; walkers may consider a drop and collect arrangement with a friendly driver.

St. John's Vale to Dockray:- 9k (6mls.)   *Allow :- 3½ hrs.*

<><><><><><>

Sector Three (continued).

Falls in the Thirlmere Valley, (Western Flanks).

The streams on the western side of Thirlmere flow from the long watershed which extends from Bleaberry Fell in the north via High Seat, Ullscarf, Greenup Edge to High Raise in the south, delineated on the OS map by the green National Trust Boundary which links these points; to the west of this watershed streams flow to the Borrowdale-Derwent catchment. The longest of the western streams and most constant in terms of flow is the Wyth Burn which empties into the southern end of the reservoir. The main streams with falls are:

- (10) Wyth Burn.
- (11) Dob Gill.
- (12) Ullscarf Gill.
- (13) Launchy Gill.
- (14) Mere Gill.

<><><><><><>

Sector Three: Falls in the Thirlmere Valley.

(10) FALLS ON WYTH BURN

Thirlmere. **NY 310117.**

The Steel End car park at the head of Thirlmere, on the tourist or back road, is a handy starting point for exploring the Wyth Burn. Across the road there is a gate on either side of the bridge; the one that provides access to the true left bank is recommended. After some 500m of level walking upstream beside the "burn" there is a stretch of rapids where the gradient increases, then a sudden narrowing of the stream where it has breached a section of harder rock to form a tight, if shallow ravine; at the upstream end of this is a charming 2½m fall where spouts of deflected water project crosswise into a long natural cistern. Overnight rain had swollen the burn and cloud breaks permitted bright sunshine to impart a brilliant sparkle to the rushing water. Further on, a bridge makes crossing easy, but before doing so look to the knick-point as far as one can see up the valley to note the distant white cascade, like residual snow on rocks.

The lower fall in the Wyth Burn.

Exploring Lakeland Waterfalls.

Across the bridge on the right bank there is a good track beside a long cascade, here tree shaded at first where cataracts of white water elbow down the rocky gradient to plunge, spread and subside. Progress upwards past where the water flows in divers ways between the scattered boulders where the bed is wide, to the final cascade first seen from the bridge, where veins of white water rush, diverge and coalesce in noisy confusion. Now the valley sides are open and treeless, clothed in bracken and rough fell grass. The approaching flow is yet gentle: here are the Wyth Burn Tarns, swellings in the stream where water lilies grow. Beyond lies The Bog.

Steel End to the top casade:- 2½k (1½ mls)   *Time:- 50 mins.*

(11) DOB GILL

Dob Gill.   **Western Flanks.**   **NY 314137.**

The water of the picturesque Harrop Tarn is carried steeply to the reservoir by Dob Gill. At the point where the gill flows beneath the road there is convenient parking and toilets at 137140. From here a forestry track ascends, initially in convenient zigzags beside the steep rocky ravine that carries the gill, although one is not immediately aware of its presence until the ridge that separates path and stream declines, to reveal on the left the delights of what at first might appear to be an artificial rock garden. Two ribbons of falling water descend a rock face on to boulders skilfully deployed, there to be shattered into a sparkling display of diamond droplets. Downstream is a succession of pools and musical cascades, set about with decorative ferns and shading trees. This is a simple masterpiece of Nature, each item of vegetation adapted to the conditions of moisture, light and shade, of yesterday's flood or tomorrow's drought. Go now and enjoy it today; explore too the upper cascade and the smooth 8m water-slide. Then continue uphill the short distance to the Tarn.

Dob Gill car park (charges) to Harrop Tarn:- ½k   *Time:- 30 mins.*

Sector Three: Falls in the Thirlmere Valley.

(12) ULLSCARF GILL
Thirlmere.	Western Flanks.	NY 313132.

The gill drains from its namesake fell of Ullscarf Edge and is the main feeder stream for Harrop Tarn. From Dob Gill car park (317140) take the track uphill to the tarn and cross the outflow by the stepping stones or the wooden footbridge. Skirt the tarn to the fence but do not continue along the obvious main track, but pass through the wire gate and continue close to the fence uphill to the stile and wooden gate. Pass over, or through, to follow the grassy track by the wire fence to the Ullscarf Gill, then cross to the other side, which is the left bank. This is the start of the ravine.

Within there is a series of short falls, interspersed with quiet pools and cascades, a charming and noisy water display. Owing to its narrowness and the overhanging birch and rowan, little can be seen of the modest falls within - but certainly heard. The steep bracken-clad slope makes progress difficult alongside the ravine in summer, but after 300m the ravine fades and the beck levels out, where it may be crossed in all but extreme conditions. Unless one intends to continue to Ullscarf, the return to Harrop Tarn is easier on the broad grassy track on the right bank of the gill. Continue round the knolls to pick up the outward path where one originally crossed the gill.

Dob Gill car park (charges) to ravine:- 1k   Time:- 40 - 50 Mins.

(13) FALLS ON LAUNCHY GILL
Thirlmere.	Western Flanks.	NY 307155.

Launchy Gill enters Thirlmere at the midpoint of the western side of the reservoir and nearly opposite Hawes How Island. There is a roadside lay-by where an information board states that the site is protected as an example of beech and oak woodland. N.W. Water (or by whatever acronym it is now currently known), Cumbria Broadleaves and English Nature manage the conservation here.

Exploring Lakeland Waterfalls.

Information leaflets and nature trail waymarkings, once offered here in the 1970's, are no longer available.

Other changes have occurred here since I began writing this and it seems steps are being taken to improve the approach and safety aspects on the right bank of the gill between the road and the wooden footbridge. As the work is on-going, every visitor must be their own explorer, but this is no playground for unsupervised children.

The Tottling stone exposed by recent felling.

Sector Three: Falls in the Thirlmere Valley.

There are two access points and two lay-bys to Launchy Gill; the easiest to describe at the present stage of reconstruction is the northern entrance. Enter through the gate and follow paths diagonally left and up the edge of the unguarded ravine where the gill slides towards Thirlmere over smooth rocks or cascades over boulders and occasional falls of 2-3m. A railed wood walkway leads to a footbridge which crosses the narrowing ravine.

The gill was visited earlier in the year following a period of exceptional rainfall. Before even reaching the walkway I was repulsed by clouds of turbulent spray rising from the ravine and saturating the atmosphere so that breathing became difficult and visibility across and into the cleft impossible. A damaged leg and smooth-soled shoes suggested that valour should be suppressed on that occasion; but to experience the power and ferocity of water on such rare occasions was a privilege.

On reaching the bridge, look upstream to the ruck of boulders that chokes the upper ravine and provides a play-way for the descending cascade. From the far end of the bridge, steps lead upward to the junction with another path. Until recently a path straight ahead led through the trees to the Tottling (or Tottering) Stone, a large boulder poised on its narrow base and perched on a hummock like a small version of the Bowder Stone. The old path has disappeared, but look ahead and upwards across recent felling to see the Stone sky-lined on its mound with just a few conifers left for company. The path to the left leads back to the road and the south entrance. The way is still rough after the felling but safer where danger spots have been fenced.

What lies upstream from the bridge on the right bank as regards permitted access is yet to be revealed by the improvements but I can tell you what is there as regards falls. The old path to the right was steep, narrow, seldom used and dangerously exposed and led to the open fellside beyond the forest boundary but was not one of the permitted paths within the plantation. It followed the right bank upstream beyond the boulder ruck to where, between sheer enclosing walls, there is a fine long fall of about 13m, broken into sections and splintered into spray by projecting shelves. Further upstream is a 10m fall fed by a 4m waterslide. Above again, the ravine widens into a ferny trough then narrows to receive another, shorter fall and water chute.

Above the boulder ruck then, are three distinct steps, each generating falls declining in magnitude as one ascends. Beyond the top fall is a grassy area where the gill flows placidly before its shattering descent. A stock fence separates the woodland from the peaty moor beyond where the gill has its origin by way of the tiny, circular Launchy Gill Tarn. This tarn should not be confused with Launchy Tarn on Dale Head. It is a coincidence that both tarns lie on the same O.S. grid line of 15: North.

From the road to the footbridge:- 250m.   *Time:- 15-20 mins.*

(14) FALL ON MERE GILL
Thirlmere.   **Western Flanks.**   **NY 299188.**

Mere Gill flows from the N-S watershed to become Shoulthwaite Gill and flow northwards beside the western edge of Thirlmere Forest. A convenient park is the loop of road bypassed by the A591 (albeit with a parking time limit) at 302205. A possible approach is to walk for about 1k along the road in the Keswick direction, past "Brackenrigg" to the fingerpost and ladder stile on the left. At this point before crossing the stile, look due east towards the hummocky fell of High Rigg. If it has recently been very wet the periodic fall of Brown Beck may be seen as a vertical white streak on the skyline, as long thin cascade of about 10m.

Now follow the footpath back behind "Brackenrigg" (calling briefly at Snipeshow Tarn perhaps) and continue beside the wall (where a pair of red squirrels sported that day), past Shoulthwaite Farm to meet the gill of the same name, now to become one's companion to its tributary, Mere Gill. Recent heavy rain had

Sector Three: Falls in the Thirlmere Valley.

transformed the stream into a series of cascades and small arching falls. Continue and cross another wall by a ladder stile. Pause here to observe the scarps high on the right hand where, in wet weather three or four more periodic watercourses descend, principal among these is a white, periodic "sourmilk gill", a cascade, prominent from the A591 and parking place. It descends from just below the skyline of Goat Crag as a narrow ribbon, disappears into a hollow and re-emerges as a white cascade.

But these are not the falls of Mere Gill which still lie over 1k onward. On the left hand is the rising woodland of Thirlmere Forest; ahead, the valley narrows to a distinct nick in the distance. Continue beside Shoulthwaite Gill to pass below the jutting brow of Iron Crag beside which another gill descends and where soon will be seen a ladder stile which crosses the forest fence on the left hand. About 150m beyond, a sudden step in the gill creates a 4m fall and the continuation of the track is made difficult by the ingress of the Mere Gill which here cuts the path as it descends from above on the right hand between low walls of rock.

The gill descends over edgewise rocks, white against grey, after negotiating an 8-10m cascade. To appreciate this, climb the steep, grassy slope to a distinct upper path. From this vantage point too, may be seen a surprise water feature, for concealed within a rock cleft above is a fine ribbon fall of 6m, deflected halfway down by a projection from the back wall.

The return might be by the outward trail or by crossing the forest fence by the ladder stile referred to above - if the gill can be forded - and following the inside of the fence northwards on a sparse but improving track until the path to Shoulthwaite Farm branches off left at 299203; from here it is but a short walk to the A591 and the car park.

Distance for the round:- 6k. (3½ mls.)   *Time:- 2 - 2½ hrs.*

◇◇◇◇◇◇

Exploring Lakeland Waterfalls.

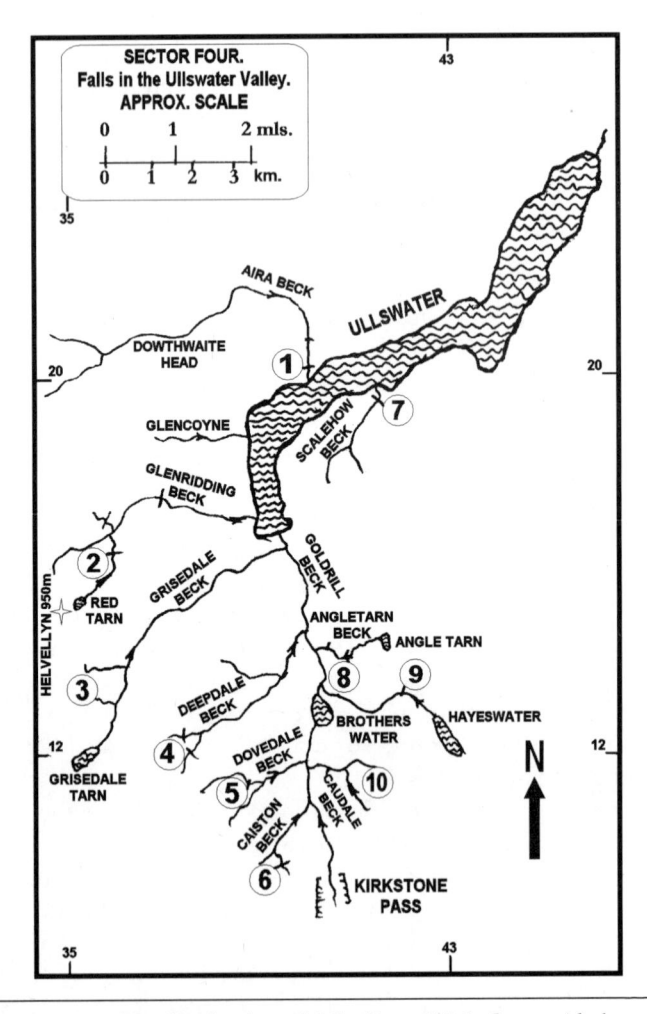

Map to show the distribution of falls (1) to (10) in Sector 4 below.

The O.S. map to cover Sector 4 is the Outdoor Leisure 5.

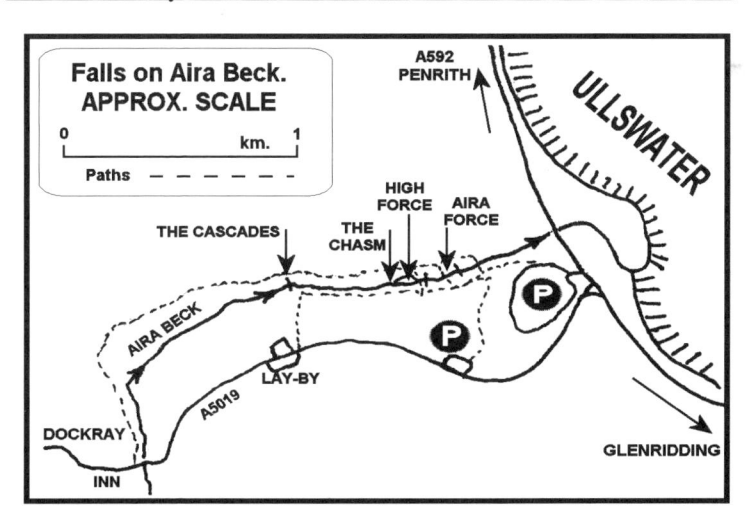

Sector Four:

Falls in the Ullswater Valley.

The Ullswater Valley extends from the Kirkstone Pass northwards to the head of the lake which then elbows north-east towards Penrith. The River Eamont drains the lake and passes tangentially south of the town. The ridges of the Helvellyn Range separate the valley from Thirlmere on the west and High Street with its Roman Road separates it from Haweswater to the east.

The main fall-bearing streams and dales are:

(1) Aira Beck.
(2) Glenridding Beck.
(3) Grisedale Beck.
(4) Deepdale.
(5) Dovedale Beck.
(6) Caiston Beck.
(7) Scalehow Beck.
(8) Angletarn Beck.
(9) Hayeswater Gill.
(10) Caudale Beck.

(1) FALLS ON AIRA BECK

Ullswater Valley.	Aira Force.	NY 400205.

Exploring Lakeland Waterfalls.

The principal fall, the easiest in terms of access from the Ullswater Valley, is the famed Aira Force, well regarded by visitors and locals alike and a favourite venue for a summer's day, easily accessible, well provided with parking, demanding only modest physical effort and with refreshments to hand.

The Aira Beck is the second longest of the streams in the valley, has the greatest catchment and therefore provides a steady yearlong flow. It rises on the northern flanks of Stybarrow Dodd, flows north via Dowthwaite Head where it bears the name of Rush Gill, aptly descriptive of this section, then circles west to Dockray and south towards the Lake.

Aira Force is a 10 or 12m series of linked cascades which culminate in a 3m fall partially concealed within a tight groove near the bottom where it debauches into the usual plunge pool. Its magnitude is enhanced by the slender, arching stone bridge which crosses far above. The enclosing walls are draped with green moss and fern near the water, while trees rise from more secure ground on either side, to lend their shade to this

Visitors gaze upon the Force from the arched bridge above the fall.

Sector Four: Falls in the Ullswater Valley.

charming feature. Secure paths enable observers to circulate around and above the Force to enjoy it from all angles. Given a wet period the cascade combines to fill the air with clouds of spray and produce, with the right angle of sunlight, the famous rainbow effect. To reach the Force, one may draw into the signed National Trust car park from the A592, and wander along the surfaced woodland walk to the fall. Another access point is from the A5091 on Park Brow at 398206 beside the road to Dockray, where there is a small car park and a steep woodland path which leads down to the Force. The distances are short, no more than 1k so allow at least an hour to enjoy the pleasures of this popular attraction. From here spectacular features lie further upstream; paths on either side of the beck may be followed but that on the true right bank is recommended else the **Chasm, High Force and the Cascades** may easily be missed. Upstream from the stone bridge over Aira Force the beck is seen below on the left hand in its dim leafy gorge until in 200m a wooden footbridge is reached; stand here and look into the deep tight chasm where the water surges some 10 - 12m below between sheer rock walls. (Note that this bridge is the last crossing of the beck before Dockray.) Just above the bridge an island splits the flow; half forms twin falls that hurtle into the chasm, the rest joins after a circuit of the islet.

The Cascades (NY 400212) lie 300m upstream from the Chasm, a feature of an entirely different nature. It consists of a broad apron fall of about 4m which drops into a turbulent pool the outlet from which is obstructed by a reef of rock two metres high where the water escapes through a narrow keyhole or vertical slit; a strange and unique feature. The top of the fall consists of water-smoothed horizontal flags mostly clear of water except in spate. To find it, continue upstream on the right bank path from the Chasm to where the best view may be had. Here the beck-side path ceases and one must take the field track to the road (where there is some free parking available) to continue to Dockray 700m

Exploring Lakeland Waterfalls.

This interesting "keyhole" waterfall may be thought deserving of a more inspired name than The Cascades.

distant. The return from Dockray is by the signed path opposite the Royal Hotel and follows the left bank downstream.

The complete round: Nat. Trust entrance to Dockray and back :- 4k. (2½ mls). Time:- 2 - 3 hrs. Parking charges apply for non-members. The round can also begin at either of the alternative free parking areas: see sketch map.

Dowthwaite Head. (NY 366210) On the section of Aira Beck known as Rush Gill is a weir, just downstream from it is a small ribbon fall of about 3m. Upstream are many charming cascades but with the cloud low on the Dodds and Matterdale Common and a steady drizzle falling, they did not inspire.

A walk to Glencoyne Head from the Aira Bridge at Dockray, across Pounder Syke then on the high path above Glencoyne Beck, across two lively gills to the exciting traverse of Glencoyne Head, returning north over Birkett Fell and beside the Coegill Beck to

Sector Four: Falls in the Ullswater Valley.

Dowthwaite Head will on a wet day, allow the aficionado to pick up a number of interesting if unimportant cascades. The waymarked track around Dowthwaite Farm is difficult to follow, but once on the field path at 371207 beside Aira Beck and the short road walk to the Inn at Dockray is over, all may seem worthwhile.

Dockray, Glencoyne, Dowthwaite Head and return to Dockray:- 11k (6mls) Time:- 4½ hrs. Map and compass essential.

(2) FALLS ON GLENRIDDING BECK

Ullswater Valley. **NY 366175.**

There are several falls on the beck but none of special significance so the initial grid reference relates to the valley Youth Hostel which is a convenient starting point for directions. This interesting valley with its long history of lead mining is worth a visit at any time whether it be in quest of waterfalls, or for its own sake and to note the relics of its former activities. There is a large pay-and-display car park at Glenridding and an adjacent Visitors' Centre where one may learn of the great flood disaster which struck the village on the night of 28th October 1927 when Kepple Cove Dam burst after a storm.

As regards waterfalls, there are no famous examples, but given plenty of wet weather some may become impressive, these occurring on the flanking streams rather than on the main Glenridding Beck. The Beck rises in Brown Cove at the head of the valley, oozing out from the base of an old dam, gathering strength from its tributaries as it goes; by the time it elbows round by the mine tips above the Youth Hostel it has become a considerable river, eventually to form a large delta which thrusts out near the head of Ullswater. From Glenridding walk up the Greenside Road beside the beck to the Youth Hostel; just beyond the first fall may be seen. Most are visible from the valley path which follows the left bank. The left, or north bank tributaries are as follows:

Exploring Lakeland Waterfalls.

The Swart Beck (365174) in spate tumbles down its rocky bed between the shattered remnants of mine buildings in a welter of foam, ending in a noisy 2m fall into a swirling pool beside the bridge before tumbling into Glenridding Beck.

Stang End Gill (361176). The fall is barely noticeable except in wet weather. Blackened rocks far up on the face of the escarpment indicate its periodic presence but at the time of passing, a fine line of white betrayed its position up on the crags.

Rowten Beck (357147) descends from a nick in the ridge beginning as a twin fall of 4m followed by a long series of cascades which streak the crags with white. This then divides into two lines of cascades which recombine before entering the main Glenridding Beck; a spectacular display, but requiring wet weather to generate its best performance.

There are no further falls on the north flank of the valley but it is worth continuing on the path upstream to see first, the damaged concrete dam that crosses the beck, then the remains of Kepple Cove Dam a little further on and Brown Cove Tarn at the head of the valley.

Red Tarn Beck at (358160) is the only fall-bearing tributary on the right bank, this can be reached by fording the Glenridding Beck just below the broken concrete dam. From here a disused leat, (water channel) makes a tolerable path to enable one to contour round towards the upper falls of the beck where there is a nice 5m apron fall with cascades above and below. Lower down and about 60m above the footbridge which crosses it, is a succession of short falls with intervening cascades, all within a shallow, sparsely-treed ravine, creating a noisy combination.

Return downstream to the remains of the mining complex on the right bank path of the main beck and cross the bridge, which spans an exciting little fall where water roars into a short ravine, is divided by a huge, wedged boulder, to burst on rocks below and empty into a calm pool above a weir.

Below the weir, the ravine continues and deepens, becoming

Sector Four: Falls in the Ullswater Valley.

dark and inaccessible where the largest fall of the main Glenridding Beck is concealed. This may be glimpsed beyond a gate signed NO ENTRY, on the side of the main track, just before approaching the first buildings.

Glenridding car park to Youth Hostel:- 2k.   *Time:- 30 mins.*
Youth Hostel to Brown Cove:- 3k. (2mls.)   *Time:- 1 hr.*

(3) FALLS IN THE GRISEDALE VALLEY

Ullswater Valley.   **Nethermost Cove Beck.**   **NY358145.**

From Ullswater, the Grisedale Valley penetrates deep into the Helvellyn Range for some 7 kilometres or 5 miles, curving gently to the south-west and flanked by great ridges; on the south by St. Sunday Crag and to the north Birkhouse Moor which terminates in Striding Edge. At its head, below Seat Sandal, is Grisedale Tarn where in the muddy bottom King Dunmail's crown is said to be buried; from here the beck flows the length of the dale to join the Goldrill Beck before their combined waters merge into Ullswater.

Parking is possible at Glenridding at the large pay and display car park, from where one may walk over the ridge into Grisedale via Lanty's Tarn (384163) by the signed path that branches south just past the shops five minutes upstream from Glenridding Bridge then descend into Grisedale to join the level track that leads up the dale. Early birds may find parking in Patterdale itself. Another possibility is the drop and collect method where Dunmail Raise or Grasmere are possible pick-up points.

If entering Grisedale from Patterdale follow the narrow tarmacked road on the south side of Grisedale Bridge, bear right and after 800m turn right just before the gate. Ahead is Brownend Plantation where the path from Lanty's Tarn is joined. The rising path leads to Striding Edge via "the Hole in the Wall" so keep to the low level track, with Grisedale Beck and Braesteads below.

Eventually the white slash of the lower fall on Nethermostcove Beck will be seen ahead as a long, swooping water slide down smooth rocks. The upper falls are 150m upstream, a steep climb

Exploring Lakeland Waterfalls.

beside the beck, hampered in summer by the dense bracken; here a 3m cascade sprays onto grey rocks and into a shallow pool.

Continue up the main valley; as it steepens keep an eye on the main Grisedale Beck where there are a couple of pretty falls; the first of 4½m is splintered into a spreading cascade beneath shading trees and a background of dark rock in a shallow ravine. See also another cascade of 3½ m further upstream.

At Ruthwaite Lodge, the climbing hut bears a plaque and dedication to the fact that it was restored in 1993 by an Ullswater Outward Bound maintenance team in memory of Richard Read and Mike Evans who lost their lives on Mount Cook in New Zealand on 31.1.1988. Behind the hut the drainage from Ruthwaite Cove and Hard Tarn descends in four prominent stepped cascades.

Just before reaching Grisedale Tarn note on the right of the path the Brothers' Parting Stone, where Wordsworth took leave for the last time from his brother John who was lost at sea two years later. Inspired by Canon Rawnsley, two stanzas from Wordsworth's Elegiac Verses were inscribed here on behalf of the Wordsworth Society.

At the tarn, curve left-about for Grasmere via Grisedale Hause and Tongue Gill; for Dunmail Raise go right, taking the Dollywagon Pike track to gain height and avoid the wet lower slopes, then branch SW and angle down to the gap in the wall where a faint path leads to

The Parting Stone where William and John bade their final farewell

the Raise Beck track which leads down to the summit of Dunmail Raise.

Allow 3 - 4 hrs. from Glenridding to Grisedale Tarn.

(4) FALLS IN DEEPDALE

Ullswater Valley.   **The Forces.**   **NY 374123.**

Where the valley road A592 crosses Deepdale Beck at 399145 there is a telephone kiosk where traditionally there has been a small pull-in used by walkers for parking; this has now been extended and tarmacked as a bus stop but still leaves room for two or three cars. Alternatively, Cow Bridge car park 1 kilometre south may be used. To enter the valley of Deepdale, take the walled lane westwards on the north side of Deepdale Bridge and at Lane Head pass through the gate on the left and follow the signed path through a second gate to Wall End. Just beyond, a clapper bridge crosses Coldcove Gill where a rush of white water from the right suggests later worthwhile exploration. Continue into Deepdale where there is easy, level walking along a narrowing track to where moraine mounds clutter the head of this broad vale. Between and over these ice-age relics, the track eases towards the right, skirting Mossydale where the drainage from Link Cove and Sleet Cove meander, to combine as Deepdale Beck and sneak past the impeding mounds down the south side of the vale.

Follow the drainage from Sleet Cove which descends a narrow groove bedecked in June with flowering rowan, by means of a steep and eroded track which needs care. The main fall at 370126 is at the head of this now considerable ravine and consists of a 7-8m plunge into a pool. Where the slope above the fall eases the stream may be crossed and lunch taken on a flat rock overlooking the spectacular drop.

Two other falls lie around the corner where the great buttress of buttresses The Step, thrusts into the dale from between Fairfield and Hart Crag to enfold Link Cove, where the next two falls are located.

To reach the foot of The Forces, traverse south-east from the lunch place at the head of the previous fall, to avoid descending the eroded ravine. Angle carefully across grassy slopes above Mart Crag and down to more level ground at the foot of the fall. The Forces is a slim, periodic fall restricted by its limited catchment; a thin stream of water glided lazily down the smooth, rounded, sloping buttress, but would produce in wet weather, a spectacular display of some 12 - 15m.

Now move 200m to the east where the main drainage from Link Cove feeds a more constant stream descending in a series of cascades, partially concealed by the narrowness of its channel and the dense greenery of its vegetation. It terminates as a braided cascade over broken rocks at the foot of the gill.

A return across the bogs of Mossydale to the firmer ground of the moraines can be negotiated but in wet weather a return along the base of the crags to the outward track is recommended.

From the A592 to Mossydale:- 3½k. (2½ mls.)   Time:- 1 hr.
Allow 1 hr. or so to explore the falls.

The short excursion up **Coldcove Gill** at 389136 on the return is pleasant on a sunny day. From the clapper bridge, take the steep grassy track left of the sheep fold and climb for about 200m, then angle right to the gill. Here the water flows down sloping slabs in a smooth, narrow water chute to a pool, from which it escapes in cheerful cascades. No prominent falls here but a pleasant setting on the tree-lined bank for the last sandwich, or cup of coffee.

(5) FALLS ON DOVEDALE BECK

Ullswater Valley. **Dove Falls.** **NY386117.**

The Dovedale Beck flows in from the west to join the Kirkstone Beck as its left bank tributary just half-a-kilometre south of Brothers Water. The drainage from Hart Crag and Dove Crag gathers in Houndshope Cove to form the Dovedale Beck which

Sector Four: Falls in the Ullswater Valley.

drops steeply for some hundred metres to form Dove Falls, a combination of waterfalls and cascades full of sound and fury in spate as it descends from the fell wall to the dale bottom.

The main downhill path from Dove Crag runs beside the beck as it descends from Houndshope Cove, the flow divides and each stream takes a separate course through the wall on the right and flows independently downhill. The path too divides, one following the water, the other continuing straight on towards Hartsop Hall. Each of the sub-streams produces falls of a scrappy nature in a lightly wooded area which can be entered by a single slab clapper bridge, that crosses the right-hand stream. Soon the waters combine and slide briskly down a worn groove in a 5-6m rock wall where a blocking boulder at the lip causes a diagonal spout to cross the main flow and rebound off the opposite wall of the groove. Below are further cascades.

The easiest approach to the falls is from the car park at Cow Bridge (404134) and by the track alongside Brothers Water to Hartsop Hall. Continue on the same line with the Dovedale Beck, which soon comes close to the path, on the left hand. Cross by the footbridge and swing right to climb beside the descending water to the falls. The return can be by the higher path just outside the fell wall as mentioned above.

Cow Bridge to Dove Falls:- 3k. *Time:- 1 hr.*

(6) FALLS ON CAISTON BECK
Ullswater Valley.　　　**Caiston Glen.**　　　**NY 393097.**

This busy beck, though lacking a significant fall, is full of variety and a delight to be beside. There are two possible approaches for the modest walker, the first is to park at the lower of the two parking areas just below the summit of the Kirkstone Pass. From there, follow the permissive path downhill and parallel with the road on the flanks of Middle Dodd. If wet, look up on the left to spot the cascades in the gill above. On reaching the Caiston Beck cross by the footbridge, climb diagonally left and pass through the fell gate to gain the path that ascends Caiston Glen. Alternatively, park at the lay-by adjacent to Caudale Bridge and walk up the pass for ½k to the signed path across the valley that links with the previously mentioned permissive path. Yet another alternative is to park at Cow Bridge (404134) and walk south on the track beside Brothers Water, past Hartsop Hall, cross the Dovedale Valley to the ancient settlement, skirt High Hartsop Dodd and curve right into Caiston Glen.

There are several cascades in the lower reaches of the beck which put on an energetic display after heavy rain but the first main feature is the long, long water slide that sweeps down a smooth, sloping channel; a dazzling sight in spate. Do not miss the point, high up on the beck, where the main stream curves in from the flanks of Red Screes. The tributary which the path continues to follow will need to be crossed to approach the steep, narrow gill which is the true source of the Caiston Beck. Within this shadowed cleft are two tiered cascades.

From here options are limited not only by the wishes and ability of the party but by the original choice of car park. If the transport awaits at the top of Kirkstone Pass, one possibility is to continue uphill on the original path to the head of the Scandale Pass and from here turn left and uphill to follow the fell wall to the summit of Red Screes, with its tiny summit tarn; *(30 mins)*. It is possible to descend from here down an airy scramble directly to the Kirkstone Inn, but be warned, one will probably arrive minus the seat of one's pants. The sight of a Chinook helicopter crossing the pass far below once prompted a direct refusal to descend from the writer's female companion. If the car is parked near Caudale Bridge one may return directly down Caiston Beck and explore the little fall on Caudale Beck. If a more robust walk is needed, turn right at Scandle Pass, take a look at the attractive Scandale Tarn and continue to Dove Crag. Try from there to locate Priest's Hole then descend the steep track to Dove Falls. Continue to Hartsop Hall and Cow Bridge or cross the valley to Sykeside and the short

Sector Four: Falls in the Ullswater Valley.

uphill walk to the parking place. *Time: (2½ hrs.).*
Kirkstone car park to Fall:- 4k. (2½ mls.) *Time:- 1½ hrs.*
Caudale Bridge to Fall:- 2½k *Time:- 1 hr.*

(7) FALLS ON SCALEHOW BECK, PLACE FELL
Ullswater Valley.   **Scalehow Force.**   **NY 414190.**

Scalehow Force lies on the eastern shore of Ullswater almost opposite Aira Point, the little delta where the Aira Beck flows into the lake. Here Scalehow Beck glides down a great buttress by a series of rounded steps to form the Force, in reality a series of cascades of elegant form involving some 30m of descent, before the beck skirts Scalehow Wood and enters Ullswater. A small additional trickle sneaks to the right of the main force as one views it from below, to form a small independent cascade. Half-a-kilometre upstream another cascade of 4½m occurs where the beck descends a backward-leaning face of knobbly rock, to produce a prominent white apron when approached from downstream.

The falls may be included as part of an excellent circular ramble to visit Place Fell. The approach to the main Force is by the pleasurable lakeside path from Patterdale, starting from the village hall at 394162 where early-bird parking is possible. Take the causeway across the Goldrill Beck to Side Farm and walk left along the path beside the lake until Scalehow Force comes into view after 3½k as one approaches the wooden footbridge over the beck. Continue for 200m beyond the bridge and take the green uphill track on the right to pass close to the head of the Force where the upper cascade shortly comes into view .

Continue southwards with the beck on the right hand and spot the second fall at 412186. By the sheepfold at the head of Low Moss Gill, a tributary of Scalehow Beck, curve right and uphill towards Hart Crag, then Place Fell with its elegant cairn.

The descent is by a clear, steep track to Boredale Hause, here

Exploring Lakeland Waterfalls.

wheel right or NW and continue downhill. At the bottom, pass through a wooden gate and immediately right through a second gate towards Side Farm and the outward route.
Distance for the round:- 11½ k. (7 mls.)　　　　　*Time:- 4 hrs.*

(8) FALLS ON ANGLETARN BECK
Ullswater Valley.　　　　　　　　　　　　　　　**NY 407140.**

Cow Bridge car park (404135) makes a convenient starting point for a modest stroll to these falls. Walk in a southerly direction along the A592 to the telephone kiosk and the left turn to Hartsop Village. Turn left again in 70m and proceed along the lane, initially tarmacked, for nearly 1k to a gate. Pass through and on to the bridge which crosses the Angletarn Beck; from here there is a good view of the main falls.

From the bridge look up the bouldery stream bed to where a single sapling has found an acceptable root hold, and beyond where twin falls of 5m drop from the lowest of three distinct rock steps. Above and beyond are lots of smaller energetic cascades which may be followed upstream if so desired, all the way to its source in Angle Tarn.

It is the lower falls which repay with most pleasure, especially if one should choose a wet day.
Distance from Cow Bridge:- 1½k.　　　　　*Time:- 40 mins.*

(9) FALLS ON HAYESWATER GILL
Ullswater Valley.　　　　　　　　　　　　　　　**NY 426128.**

Hayeswater Beck, the outflow from the tarn of the same name, now dammed to provide a local water supply, may offer noisy but modest entertainment for those seeking further exercise.

From the true left bank of the Angletarn Beck, midway between the footbridge and lower falls is a ladder stile from where a path leads south to Hartsop. Here, at the head of the hamlet is a small car park which offers an alternative to Cow Bridge. Pass through

Sector Four: Falls in the Ullswater Valley.

the gate by the sheep pens and proceed uphill. Soon the track divides. The left tarmacked road leads to the Filter House where a steep path descends to a footbridge over the Hayeswater Gill and rises to the left bank track. The alternative is to continue past the junction with the tarmacked road, descend to the gill and cross by the wooden footbridge and climb the rocky road to the point where it meets the previously described track. A short distance upstream from here the main fall occurs but one must descend the steep side of the gill to view it. Squeezed between narrow walls, the water descends in a two-step cascade of thundering water; Take care on the slippery bank not to join it. Continue now, the main track regained, to the Tarn.

From Hartsop to Hayeswter:- 2k.   *Time:- 50 mins.*

(10) FALLS ON CAUDALE BECK
Ullswater Valley.   **Kirkstone Pass.**   **NY 404115.**

There are a number of small lay-bys beside the A952 on the approach to Kirkstone summit, the handiest and most convenient in this instance being at Caudale Bridge (402115). Alternatively there are two car parks near Kirkstone summit from where one may walk downhill on the permissive path as if going to Caiston Glen but instead crossing the Kirkstone Beck to the roadside gate at 402111 and continuing down the road to Caudale Bridge.

From the bridge walk 75m north to a finger post and stile indicating the access point to the fell on the eastern side of the road. Climb diagonally right until the path becomes clearly discernible and passes through bracken to a ford. Immediately below the ford is the main fall of 3½m where a curtain of sparkling water leaps a knuckle of smooth, rounded rock then descends energetically as a series of cascades to the road bridge. Just above the ford twin jets emerge from a dark, vegetated cavity. Ford the crossing and continue uphill on the left bank on the narrower track which follows close to the beck, to where there is a pleasant twin

Exploring Lakeland Waterfalls.

water slide of about 12m. Above, the water can be seen leaping from step to step in a long series of cascades; berried rowan stand bank-side at intervals. The main Caudale Beck approaches from the right hand as one ascends and where there are two series of cascades; the water flowing from straight ahead is in fact a tributary stream. This is a pleasant water to enjoy on a warm summer's day.
Distance from the road to the ford:- under 1k.   *Time:- 10 mins.*

Fall near summit of Kirkstone Pass (404091). Don't miss the delightful little cascade seen as one drives up the pass from the Ullswater side towards the summit, where one's passengers may glimpse this intriguing little feature on the left, just across the roadside wall.

Best to stop at the lower of the two car parks just before reaching the summit, cross the A942 and walk downhill for 150m. to view it. Beware the holiday traffic.

◇◇◇◇◇◇

Sector Five:

Haweswater and Swindale.

The Haweswater Valley - or Mardale as it once was - lies to the south-west of Ullswater and is separated from it by the long ridge of High Street. The dam was constructed during the 1930's, from the first borings for the dam being taken in 1929 to the final filling of the reservoir in 1941. The purpose of Manchester Corporation was to supplement the water supply to the industrial north-west. In the process, the village of Mardale at the head of the valley was submerged, occasionally to reappear in times of drought when the outlines of buildings and walled lanes may be traced.

The falls in this area are modest in height and mostly in the form of cascades or gills; the adjacent valley of Swindale included here, displays the best falls in this sector. The fall-bearing streams are as detailed below.

(1) Howe Beck.
(2) Measand Beck.
(3) Haweswater Beck.
(4) Blea and Small Water.
(5) Rowantreethwaite and Hopgill Becks.
(6) Swindale Beck.

The map to cover this sector is the OS Outdoor Leisure 7.

Exploring Lakeland Waterfalls

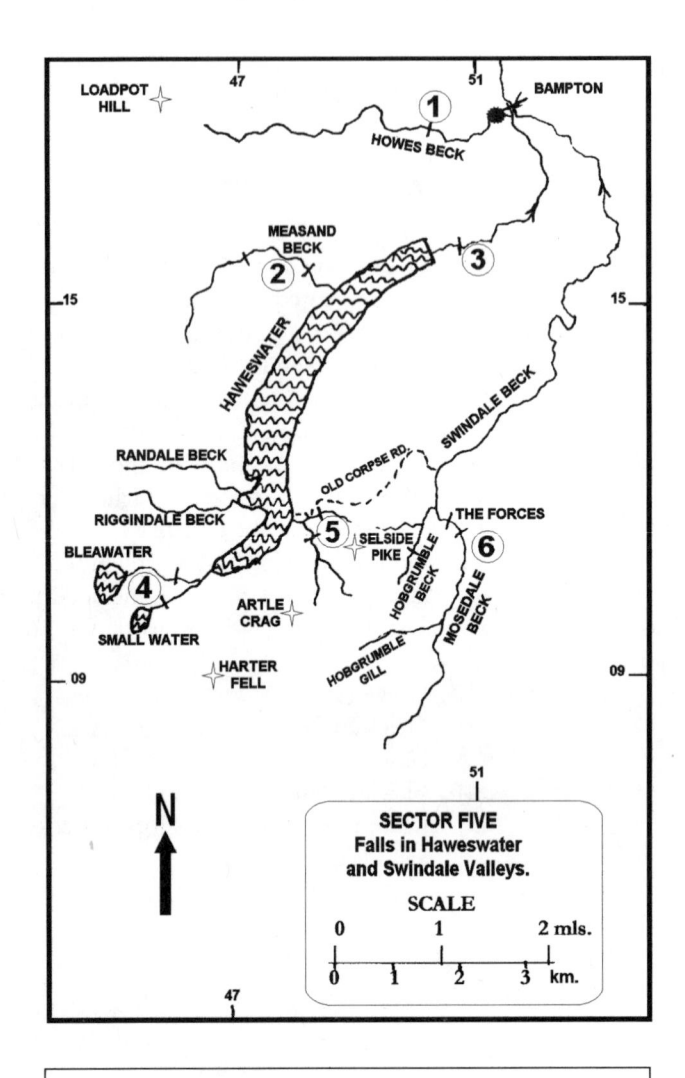

Map to show the location of falls (1) to (6) in Sector 5.

Sector Five: Haweswater and Swindale.

(1) FALL ON HOWE BECK

Haweswater.	Force Hole.	NY 501177.

The fall lies 1½k. due north of the Haweswater dam; the beck gathers its water from around Loadpot Hill and flows east as the Cawdale Beck, then as the Howe Beck, to Bampton. There is not as far as I am aware, a public right of way to the fall beyond the small dam on the Howe Beck (at 504177); but a visit was made to it on a wet April morning when the water was in full flow. The lack of available parking on the narrow roads around Bampton made the approach difficult; this was resolved by visiting a convenient campsite in the village. It may also be possible to find parking on the west side of the bridge that spans the Haweswater Beck By Walmgate Foot (518176).

Force Hole Waterfall on the Howe Beck.

Head for Bampton and just north of the bridge that crosses the Howe Beck turn left and follow the sign for Hullockhowe, noting in passing the rapids at Millcrags. Keep left at the hairpin and left again onto the road to the Howes where gorse predominates; here one comes close to the Howe Beck where there is the small dam mentioned above, and a footbridge. Do not cross but keep to the true left bank of the beck; continue through a field gate and skirt the small waterside plantation for about 300m. The fall will be heard through the trees on the left hand.

The Force drops in two beck-wide steps, the upper introductory fall being of 1½ m. in height, the lower a broad apron of 2½ m. in width, falling 3m. into a shallow rock gorge. The peat-brown water, streaked with foam, glided away downstream beneath trees which reached across from bank to bank.

Walking distance from Bampton Bridge:- 2½k.   *Time:- 50 mins.*

(2) FALLS ON THE MEASAND BECK
Haweswater.   The Forces and Fordingdale Force.   NY486155 etc.

The glory of the Measand Beck is hard to beat in Lakeland for sheer beauty and variety, particularly the section between the lake shore and the upstream footbridge at 483156. For those who have been fortunate to visit the water gardens of Monesteria del Piedro in Spain, created by the Spanish monks in the twelfth century, this is a natural version in miniature.

If one approaches along the shore path via the gated bridge, cascades may be glimpsed below where the water tumbles into the lake, but above, tracks partly concealed in summer by bracken, provide access to the more impressive features where, about 40m uphill from the lower bridge, the braided beck displays 4 or 5 separate falls and cascades of differing forms, to produce a variety of sound and scenery to bring delight. Nowhere does any drop exceed 5m and all coalesce within a rock basin overhung by ash and willow. High above, from a prominent rock buttress decked with heather and shaded by trees, triple cascades descend. Around the corner in secret crevices, thin streams of water make a temporary escape. Higher yet, narrowed by confining walls

Sector Five: Haweswater and Swindale.

Fordingdale Force on the upper Measand Beck.

colourful with purple heather and mottled greens of lichen, the channelled waters roar in mini-falls and cascades.

As the bridge upstream is approached the slope of the waterway decreases and in Fordingdale Bottom the flow is yet smooth and gentle before its impending plunge.

At the footbridge, choose the side of the Measand Beck you would prefer to follow in order to avoid the boggy area ahead. Approaching Force Crag the tranquil beck becomes more animated where a pair of elongated water slides descend smooth, sloping slabs. Then follows a most delightful water feature; a knuckle of rock, thrust up in the centre of the watercourse and surmounted by a squat rowan, divides the flow into two chattering cascades; on a day of brilliant blue and summer sun a sight to charm the eye. This is **Fordingdale Force,** (471158) modest though it may be. In a shallow rocky defile upstream chuckling cascades and sublime rock pools proliferate.

Exploring Lakeland Waterfalls.

Seven hundred metres upstream a series of shallow, stepped falls occur preceded by a 3m cascade over rounded, mossy rocks, then just beyond a broad-boughed rowan tree is a 3-4m twin fall (463156) down a blackened rock face; just a trickle on this summer's day.

Car park to Measand Beck:- 2½k   *Time:- 45 mins.*
Measand Beck to Fordingdale Force :- 1k   *Time:- 30 mins.*
Distance to final fall:- 700m   *Total time:- 1½ hrs.*

(Park as for Thornthwaite Force, below.)
From here, at the final fall the choice is to return or to proceed upstream to Low or High Raise or Kidsty Pike with a steep return to the reservoir path at Bowderthwaite Bridge. A descent of the Randale Beck to inspect the waterfalls marked on the OS map is not recommended, for the reality does little to compensate for the ankle-cracking exercise where, despite the indication of a path, a multitude of irregular rocks disguised by grass and moss, wait to ambush the unwary.

(3) FALL ON HAWESWATER BECK
Haweswater.   **Thornthwaite Force.**   **NY 513161.**

The Haweswater Beck flows from the base of the dam which retains the Haweswater Reservoir and passes beneath Naddle Bridge where the road crosses to run beside the lake to Mardale Head.

The force is found 300m downstream from the old stone bridge which crosses parallel to the new road bridge and a few metres downstream from it. Parking is possible in the Burnbanks Cottages area; turn right at Naddle Gate and drive along the lane to where there is usually room on the left opposite the cottages.

Walk to the bridge and cross the ladder stile on the north side then cross the grass-surfaced stone bridge to the permissive path, part of the Cumbria Way, and follow it downstream to a wall and another ladder stile, from where the fall may be seen.

From the river bank enjoy the sight and sound of the water,

Sector Five: Haweswater and Swindale.

Thornthwaite Force on Haweswater Beck.

first the introductory cascades then a convex fall of 4-5m, white against the surrounding trees, edging bushes, and bouldery flanks. A broad pool partially filled with shattered stone receives the water which flows away north-east towards Penrith.

Distance from parking place:- under 1k   Time:- 20 - 30 mins.

(4) FALLS ON BLEA AND SMALL WATER BECKS
Haweswater.   NY 460107 and NY 460104.

At Mardale Head at the head of Haweswater there is plenty of parking space, albeit much contested in summer: a popular springboard for the magnificent walking in the area. The falls are modest and should be visited not simply for themselves but in combination with the associated tarns and summits. The two becks, as their names suggest, flow from their respective tarns.

From the parking area take the Nan Bield Pass track keeping an eye to the right for the first and most interesting water feature, which occurs on the Blea Water Beck. Head towards it across wet and hummocky ground and cross the Small Water Beck. If conditions are too wet to cross, it may be better to turn right at the fell gate and cross the Mardale Beck (which the combined Small Water and Blea Water Becks have now become) and walk upstream to its confluence with Blea Water Beck.

The fall consists of a drop of nearly 4m onto broken rocks within a deep cleft; two birches stand prominently at its head while berried rowan flank the sides. A slanting water slide can be seen above.

Return to the main Nan Bield track through the gate in the intake wall; the path and Small Water Beck gradually converge and another water slide can be viewed on the right bank. A further 100m uphill the final feature of interest occurs where a cascade descends 6-7m down shattered rocks. Enjoy the rest of the ascent to the tarn where stepping stones permit an easy crossing of the outflow.

Parking place to Small Water:- 1½k   *Time:- 30 - 40 mins.*

(5) FALLS ON ROWANTREETHWAITE AND HOPGILL BECKS

Haweswater.　　　　　　　　　　　　　**NY 484117 and 483116.**

Beside the Old Corpse Road that ran from Mardale to Shap where the deceased were once carried for burial, are two active becks that in wet weather descend white with fury, combine and enter Haweswater. If passing that way, not accompanied by mourners, they should be visited. From beside the reservoir the steep gradient of the Old Road with its torturous zigzag bends may be better enjoyed downwards if combined with other delights; an ascent perhaps of Harter Fell from Mardale Head by way of Small Water and Nan Bield, returning north-east by Artlecrag Pike and Selside Pike to meet the Old Road. Gatescarth Pass provides a shorter

Sector Five: Haweswater and Swindale.

round, but if time presses, the direct assault from the reservoir will take but 30-45 minutes.

When descending from the longer rounds, the headwaters of the Rowantreethwaite Beck are first crossed as one descends the Corpse Road, with the watercourse on the left hand, but it is not until lower down and at the bends, that a good view back up the beck may be had and where its flow has been increased by the capture of three other unnamed gills. Enjoy in wet weather, the sight of its exuberance as it bounces down its stepped, incised little valley, at least to the point where screening trees have clustered to take advantage of the shelter within its protective flanks.

Lower still at the penultimate bend, a view up the lower Hopgill Beck reveals a similarly excitable watercourse, although more exposed and rugged with fewer trees. Both becks join out of sight below a convex slope and pass as one beneath the metalled road below, to enter the lake.

From Mardale Head car park:-
The longest round (via Small Water), distance:- 11k   Time:- 4 hrs.
The distance via Gatescarth Pass:- 8k   Time:- 2¾ hrs.

(6) FALLS ON SWINDALE BECK

Haweswater.   Forces Falls.   NY 509115.

The Swindale Valley lies parallel to the Mardale Valley or Haweswater Reservoir over the ridge to the east of it. The Old Corpse Road between Mardale Head and Swindale Head links the two valleys. A practical route to Swindale by car is through Bampton and south-eastwards through Bomby, Rawfoot to Toathmain where an unfenced road branches right. Off road parking is possible in the region of the Filter House (524146); there is no parking higher up the valley.

Walk along the valley road beside the Swindale Beck to Swindale Head and continue along the track, cross the footbridge to Dodd Bottom and beside what is here the Mosedale Beck. From the Geordie Greathead Crag on the right hand a fearsome tree-flanked cleft splits the fellside, channelling the drainage from above, between the domed moraines of Dodd Bottom, to join the Mosedale Beck. Viewed from below there are three linked falls each enclosed between sheer walls, the middle one being the greatest at 10 - 12m, though the falls within are reduced to a trickle in dry weather. In wet conditions this is a grim inhospitable place well deserving its sinister name: **Hobgrumble Gill,** (502114).

Forces Falls are the upper pair of water features (section 5) of what is really a long succession of cascades on the Mosedale Beck, which extend for a distance of almost half-a-kilometre during their 120m of descent. From the footbridge and proceeding upstream the falls may be described in sections as follows.

1) Twin cascades split by a boulder descend for 5m, then 1½m creating prominent white waterfalls as the beck is first approached. Note the constructed stone erection on the ledge on the right bank which may have been the base for a former, or projected bridge prior to the building of the present one 20m below.

2) The above initial cascade is powered from a deep, calm pool which in turn is fed by a pair of low-angled cascades of 5m and 2m respectively.

3) The beck steepens and becomes more turbulent where a cascade of three steps descends beside a splinter stream which has found a parallel course. An eddy scoops, here out of the main flow, dark peaty water topped with tan foam stands like stale Guinness in uncollected glasses.

4) Another series of noisy cascades follow, above which the beck is squeezed between narrowing walls; higher yet a ruck of jammed boulders break the flow into 4 or 5 separate streams of white, foaming water.

5) The downward slope increases, the banks become more precipitous and the increasing sound of falling water introduces a

Sector Five: Haweswater and Swindale.

5-6m steeply angled water chute which swoops into a deep brown pool. Above the slide, an elongated plunge pool accepts a roaring cascade that thunders down a steep narrow cleft. Beyond the lip of this final feature the water approaches calmly, as yet unaware of its impending aggressive mode as it meanders through the upland vale towards the downfall. Viewed in a wet summer, Forces Fall impressed itself as one of the finest succession of cascades in Lakeland.

A return may be made over Nabs Moor towards Selside Pike then north to the Old Corpse Road and down to Swindale Head with the return walk along Swindale Lane.

From the Filter House to top of the falls:- 5k *Time:- 1 - 2 hrs.*
Return via Selside Pike and Corpse Road:- 7k *Time:- 3 hrs.*

<><><><><><>

Sector Six:

Longsleddale and Kentmere.

These charming valleys lie in the south-eastern fells of Lakeland, their remoteness emphasized by difficult access along single-track roads and a lack of parking facilities. This has gone some way towards preserving the rural way of life against the flood of tourism and the demand for second homes that has overwhelmed other communities in more accessible valleys.

The main entry to Longsleddale is by way of Garnett Bridge (SD523993) with access from the A6 north of Kendal or from the A591 through Burneside. The normal limit of motor access is Sadgill beside the River Sprint; here there is limited parking. Beyond this point the rough road struggles north to the Gatescarth Pass then descends to the Haweswater reservoir, the pass has been used as a "green road" by motorcyclists and off-road vehicles causing extensive erosion. In this upper section of the dale the steep eastern flanks have been rent by deep, narrow gills that hold the prospect of exciting water-features in times of excessive rainfall. In summer they are virtually dry.

The OS map to cover this Sector is the Outdoor Leisure 7.

(1) FALLS ON THE RIVER SPRINT
Longsleddale.   Wren Gill.   NY 477078 & 478085.

Strolling northwards from Sadgill, the first feature of interest is the low stone barrage that crosses the valley at 481068 to form a shallow pool. Upstream where the gradient is steeper, the white water of cascades can be seen contained within a trough of smooth blue-grey rock where it produces a series of pools and falls: a delight for eye and ear on a summer day. Further upstream the thickly wooded gorge of Cleft Gill precludes vision; here secret waters roar. Higher up the road at the point where it bends

Sector Six: Longsleddale and Kentmere.

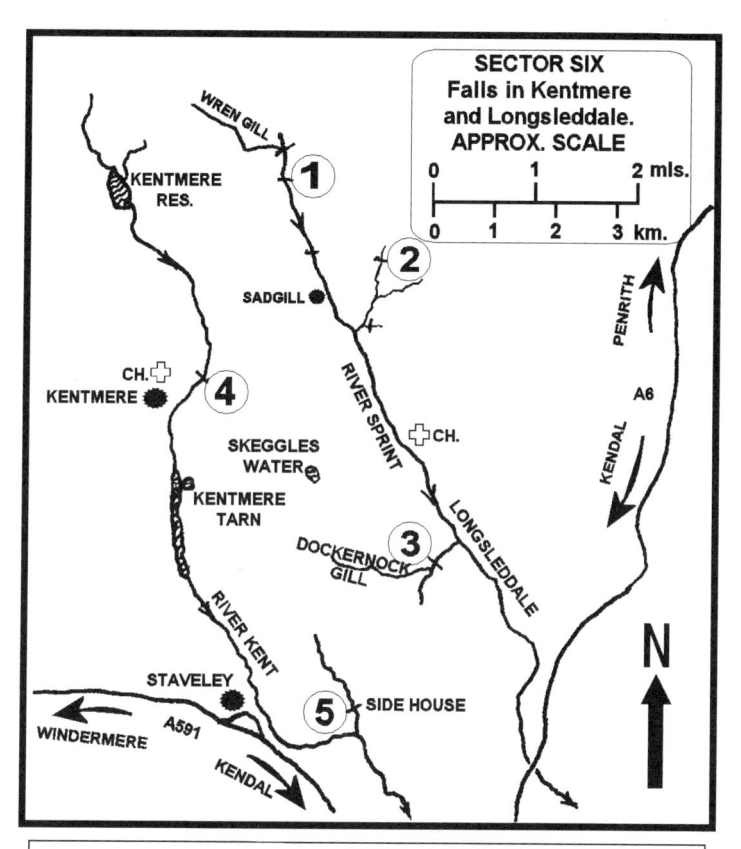

Sketch map of Sector 6 showing the location of falls (1) to (5).

sharply right then left a triple-tongued cascade bends over a mid-beck buttress with a brisk "whoosh" and a scattering of sparkling spray.

The repaired road surface of edgewise slate, a real bone-shaker for mountain bikes, leads steeply to the head of the pass. Here Wren Gill hurries in from the left hand to introduce with a series of white spouts the best fall of all on this beck, a thin ribbon fall of 2m, then 4m, drops between recessed rock shoulders. Here the River Sprint is born.

Sadgill to head of the Gatescarth Pass:- 3k. *Time:- 1 hr.*

(2) CASCADES NEAR STOCKDALE BRIDGE
Longsleddale. **River Sprint.** **NY 493057.**

While driving up the dale towards Sadgill the eye was caught by a series of cascades on the hillside east of the road at Stockdale Bridge. A lucky parking space nearby saved the trail back from Sadgill. From the bridge a rough road leads east in the direction of the cascades, past a cottage, an Outdoor Pursuits Centre, some disused quarries, an interesting and well preserved lime kiln, to a densely wooded copse where, in the dim green of a rocky gorge the Brow Gill creates a 3m fall between two series of stepped cascades. A railed enclosure nearby protects the bridge across the gill that carries the water pipeline from Haweswater southwards. Beyond a snecked gate a delightful series of seven cascades came a'leaping down a steep green fellside and which had first attracted attention from the dale road. These occur on an unnamed gill west of Brow Gill.

It was this chance exploration that led to the discovery of the Longsleddale Outdoor Pursuits Centre and its work with Rangers, Scouts, Police Cadets and School Groups. It runs on a shoestring aided by voluntary donations, from basic but adequate premises after its move up the dale from what is is now the St. Mary's Church Community Centre; *(see Dockernook Waterfall)*. It runs

Sector Six: Longsleddale and Kentmere.

courses on general outdoor activities, from Duke of Edinburgh Award Expeditions to basic rock climbing, fell craft, geography and biology. The Centre leans modestly towards Christian principles and exudes a relaxed, friendly, hospitable atmosphere: this impression was reinforced by the offer of delicious bacon and eggs.
Stockdale Bridge to cascades:- under 1k.   Time:- 20 mins.

(3) DOCKERNOOK WATERFALL

Longsleddale.   **Dockernook Gill.**   **NY 498009.**

St. Mary's Church of Longsleddale is set on a prominence at the midpoint of the dale. Adjacent is the Community Centre, once the the schoolhouse which fell into disrepair when pupil numbers waned and busing took over. It was restored in the late 1960's as the Longsleddale Outdoor Pursuits Centre, but by mutual agreement this moved further up the dale to a more convenient location in 1973. A public toilet maintained by the Ladies of the Parish (bless them) is sited beside the Community Centre; here too is the only official parking area in the Dale. The parking and facilities have been created by the local community for the local community and with this in mind should be treated with respect, especially on Sunday mornings.

From this car park a brisk twenty minute road walk south will bring one to the signed bridleway to Dockernook (509016). Walk up the long approach to the farm, pass through the two gates beside the farmhouse and zigzag up the hill following the faint track. At the intake wall use the gate beside the ruined building. In 300m a tiny beck crosses the track, follow this to the left and downstream to Dockernook Gill, passing through the gap the little stream has eroded in the mound of moraine.

The fall is a pretty feature, sheer and of nearly three metres in height, where a resistant lip of limestone crosses at a slight angle to the flow: here a large rowan dominates. At the water's edge a

Exploring Lakeland Waterfalls.

broad limestone slab forms a fitting dais from which to enjoy the spectacle.

Distance from the road:- 1½k *Time:- 30 mins.*

Dockernook Waterfall, a small isolated feature of much charm.

Sector Six: Longsleddale and Kentmere.

Kentmere.

Kentmere, with its narrow approach road with passing places, is also notoriously difficult of access due to its very limited parking. Beside the Church and Community Centre is normally the only available parking area and instructions suggest when and where parking is permissive. In order to alleviate the difficulties caused by an influx into a small farming community of numerous extra vehicles, the camping field beside Low Bridge as the village is approached is sometimes available for parking in summer.

(4) FALL ON RIVER KENT		
Kentmere.	**Force Jump.**	**NY 461044.**

To reach Force Jump, the easiest approach is to take the steep metalled lane opposite the occasional parking field beside Low Bridge to the sharp bend at the top and turn left into High Lane. About 220 metres from the corner, the walled but ungated Low Lane dips off to the left, where a gap in the wall gives access into an area of rough woodland which slopes down to the River Kent.

From the beck side looking upstream, the water appears to approach over a series of natural weirs formed by lines of boulders or irregularities in the stream bed that cause the flow to approach in a series of leaps and jumps. A small wooded island divides the flow but on reaching the downstream corner of the islet it re-unites and enters a short ravine. Here the total flow of the Kent is compressed through a gap a little more than a metre wide; on the occasion of the visit it dropped steeply for 4m in a boiling fury of white, above which hovered a fine mist of spray held by the slanting winter sunlight. After a 6m surge between the moss-lined walls, the water dipped steeply, to be deflected by some submerged obstruction causing it to arch over and down for a further 2m into a smooth, widening channel where its energy was released in rising bubbles of air like a natural jacuzzi, once popular with the local

Exploring Lakeland Waterfalls.

youth as a summer bathing pool .
    Distance from the Church:- 1¼k   *Time:- 15 - 20 mins.*

After gaining access to Kentmere it would be wasteful not to explore some of the many paths in the area. For fall hunters an obvious choice might be to continue northwards along Low Lane, past Overend and onwards to Kentmere Reservoir at the head of the valley, a long but fairly level walk, to check on the small but energetic gill-like falls to be found on the upper Kent and Lingmell Gill.

(5) FALL NEAR SPRING HAGG, STAVELEY

Kentmere Valley.   **Fall by Side House.**   **SD 489986.**

From Staveley take the road north towards Kentmere and shortly cross the River Kent at Barley Bridge, to turn sharp right onto the narrow lane which eventually becomes the Potter Fell Road. However, after just a kilometre take the first turning left and uphill to Craggy Plantation where there is parking for about three vehicles. Walk back down the hill, turn left at the junction and continue on along the road to the first finger-post at 480983 just before a double bend sign. Cross the stone stile and walk on the faint path across the field, to bear right at Dorothy Farrar's Spring Wood and onto the woodland track to the Spring Hagg kennels where the way becomes metalled. Continue on the same line to the junction, there turn left and shortly through the gate to Side House, where it is easy to locate the fall with ear and eye.

Except after heavy rain the flow of the unnamed beck is small and cascades down a series of green-algaed limestone steps for about 3½m. It is easy to reach the base of the fall, but in summer the area is heavily shaded under tall trees.

Return either by the outward path or alternatively retrace one's steps as far as Spring Hagg Kennels but take the right hand path which leads northwards along a pleasant walled lane to the

Sector Six: Longsleddale and Kentmere.

unfenced road near Piked How, where a left turn will bring one back to Craggy Plantation.

Other alternative return routes through Frost Hole or Birk Field would seem to be insensitively routed through private property and out of consideration for the occupiers are not recommended. Do explore Craggy Plantation before leaving the area.

Craggy Plantation to Side House:- 2k. *Time:- 20 - 30 mins.*
Recommended return distance:- 3k. *Time:- 30 - 40 mins.*

A small, somewhat unatttractive waterfall downstream from Frost Hole and beside Side House.

Sector Seven:

Falls in the Windermere Catchment.

Windermere, the largest of the Cumbrian lakes, accepts the drainage from an area greater than any of the other Lakeland waters and mainly from the north and west where most of the fall bearing streams occur. Here we are looking at Grasmere and Rydal and the rivers that flow into these smaller lakes then enter Windermere via the River Rothey. Then there is the considerable catchment of Great and Little Langdale which merge to become the River Brathay; this joins with the Rothay just before entering Windermere. The only other stream of interest is Mill Gill which enters from the east and a tiny flow that comes in via Esthwaite Water.

The main falls or fall-bearing waters are as follows.

Rothay Waters:
(1) Tongue Gill.
(2) Green Burn.
(3) Greenhead Gill.
(4) Blindtarn Gill.
(5) Wray Gill.
(6) Sourmilk Gill.
(7) High Fall, Rydal Beck
(8) Stockgill Force.

Brathay Waters:
(9) Skelwith Force.
(10) Colwith Force.
(11) Thrang Crag, Meg's Gill.
(12) Dungeon Ghyll, Stickle Gill.
(13) Whorneyside Force.
(14) Browney Gill.

Others are: (15) Thurs Gill (on Hawkshead Hill), (16) Mill Beck (for Rayrigg Cascades).

The OS maps to cover this area are Outdoor Leisure 6 and 7 (and 5 to cover Tongue Gill Falls). The sketch maps to show the location of the falls are on pages 107 and 128.

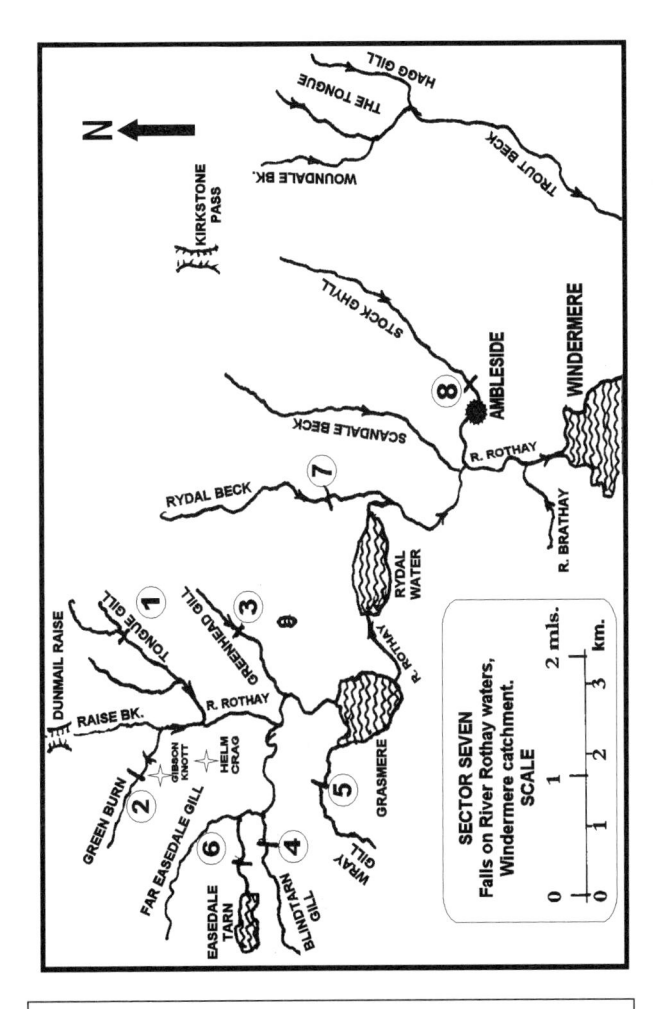

Sector Seven: Falls in the Windermere Catchment.

Map to show the location of falls (1) to (8) on R. Rothay waters. For map of Brathay streams see pg. 128.

Exploring Lakeland Waterfalls

(1) TONGUE GILL, GRASMERE

Windermere. **R. Rothay.** **NY 348111.**

There is ample parking in the Grasmere area either for free or at a price, depending on availability. There is even parking, if one is early enough, on the corner of the side road opposite the signed path to Tongue Gill by Mill Bridge (336093) on the A591 just north of the Travellers Rest. Walk NE beside Tongue Gill, through the gate and continue for 100m where there is a narrowing of its entrenched valley. Here is concealed Tongue Gill Force, but there is no suitable path to it and the precipitous sides are unstable and dangerous, emphasised by a partially vandalised notice. Further upstream one may look back to see many collapsed trees which have fallen into the beck to confirm the necessity for caution.

Continue to the second gate and cross Little Tongue Gill by the giant stepping stones, then the main beck by the timber footbridge, to continue upstream where ahead, the two white snakes of the upper falls will be seen at the head of the valley. The path passes between them at 348111.

The upper fall steps down nearly vertical strata, blackened by moss, in a dozen silver threads which weave patterns and glisten in the afternoon sun; under wet conditions this presents a totally different aspect of combined boisterousness. The height of the upper fall is about 10m. Below the path, the flow continues to a three-stage cascade with drops of between 3 - 5m. Beyond, a steep constricted chute rushes to the valley floor.

The two becks, Tongue Gill and Little Tongue Gill are separated by the elongated fell of Great Tongue. From the falls, one may, with little extra effort, slip over the ridge and descend for a change by Little Tongue Gill, or indeed along the ridge itself. This is also a good approach to Seat Sandal or Fairfield.

Distance, Mill Bridge to Upper Falls:- 2½k. *Time:- 1 hour.*

Sector Seven: Falls in the Windermere Catchment.

(2) FALLS ON GREEN BURN, GRASMERE

Windermere.   **R. Rothay.**   **NY 325101 & 321104.**

There is plenty of car parking space in Grasmere village but in order to shorten the walking distance park as for Tongue Gill north of the Travellers Rest or in the lay-by at the bottom of Dunmail Raise. From Mill Bridge walk down the lane to the Low Mill bridge; turn right or north to follow the R. Rothay upstream to the confluence of Raise Beck and the Green Burn at Ghyll Foot. Keep straight on past "Helmside" where the path now follows the Green Burn to the footbridge; just below is the first fall of just 2½ m. Continue along the path on the left bank. Ahead may be seen a prominent breast of rock rearing up at a step in the valley. White cascades vein its rugged 10m face. The track skirts around to the right of the rock face and connects with the stream above the fall which here flows in a shallow rock channel. Beyond, cheery cascades occur until one enters an extensive basin of marsh and moraine. The track swings left and crosses the burn by substantial stepping stones.

Distance to this point:- 2½k.   *Time:- 1½ hours.*

From the stepping stones two alternatives are offered.

1). Continue onto the ridge at Calf Crag or Pike of Carrs and follow the ridge in a south-easterly direction past Monument Crag to Gibson Knot, across Bracken Hause and onto Helm Crag to pat the Lamb and feed the Lion. Return to Bracken Hause and descend the grassy, slippery slope to join the outward path at Ghyll Foot. Where the path from Bracken Hause reaches the Green Burn there is a two-step cascade of 5m and 4m, which may be heard but not seen as it lies within the private grounds of "Helmside".

2). From the ridge at Calf Crag descend SW into the Upper Easdale Valley at Moor Moss, a low-slung hanging valley and head downstream. This is a strange other-worldly place where side streams percolate into the main flow through mini-gorges, short tunnels and other unusual features. A black-walled cascade rushes

Exploring Lakeland Waterfalls

down and further on another, topped by holly bushes, into a short ravine; all features fun to explore. A sheepfold follows, then Stythwaite Steps, but use the footbridge just downstream to cross the Far Easdale Gill. From here it is a goodly walk back to Goody Bridge where a sharp turn left leads one to Low Mill Bridge; finally, right to Mill Bridge.

1st alternative; from stepping stones via Helm Crag:- 7½k.   Time:- 4 hours.
2nd alternative; via Far Easdale valley:- 9k.   Time:- 4-5 hours.

(3) FALL ON GREENHEAD GILL, GRASMERE
Windermere.   R.Rothay.   NY 351089.

The Swan Hotel is situated at the cross roads on the main A591 where at 339082 one road leads west to Grasmere village, the other minor road east to hotels and residences. Take the minor road east to where it splits into two loops left and right, which circle back to the main road. Take the northern or left loop to where, in about 100m a walled, tarmacked path signed "To Greenhead Gill and Alcock Tarn" will lead to a gate. This in turn is labelled "Public Footpath to Stone Arthur and Alcock Tarn".

Cross Greenhead Gill by the footbridge and ascend the steep path ahead which soon doubles back to a grassy terrace overlooking the gill. Here is a seat dedicated to Tennyson Oldfield (1892-1978), author of "Come for a Walk with Me". In the gill below is the aqueduct that carries the Thirlmere to Manchester water pipeline. Continue along the track with the walled woodland on the right hand; where the wood ends the main track swings right and uphill towards Alcock Tarn and Heron Pike. Below is a small private reservoir, the outfall from which gives rise to a merry cascade.

Continue along the lesser path which clings to the hillside above the gill as it curves to the left where, some distance ahead, the white water of the falls lures one onwards. The valley narrows and as the falls are approached they resolve into three separate features.

Sector Seven: Falls in the Windermere Catchment.

The lower, partially screened by a spreading rowan, is seen to be a 2m ribbon of descending water. Beyond are two long, slanting water slides, one above the other. Nothing spectacular here, but a pleasant morning's stroll.

Distance:- 1½k.   *Time:- under 1 hour.*

To complete the round via the summit of Stone Arthur, angle up the slope above the falls in a NW direction until the craggy rocks of the summit come into view; continue on the bearing to pick up the path to the cairn. (A continuation along the upward path would bring one to the summit of Great Rigg.) To return to the day's starting point, follow the track downwards; where it crosses a small gill a fall can be heard and glimpsed. After this the well used path becomes very steep and needs care but will return one to the start of the walk.

Distance; Falls to Stone Arthur:- ½k. (25m of ascent)   Time:- 45mins.

(4) BLINDTARN GILL GRASMERE

Windermere.   **R. Rothay.**   **NY 322080.**

From the centre of Grasmere walk along Easedale Road, cross the Easdale Beck at Goody Bridge and past the road junction on the right that leads to Mill Bridge, then in 200 metres spot the signpost that points the way left to Upper Easdale. Cross the two footbridges and walk along the stony beckside path across the field. Just beyond the New Bridge note the gate on the left with a yellow waymark. Pass through, cross two fields separated by a short section of woodland to the farm and cottage and continue to the gate which has a waymark on the uphill side. Blindtarn Gill is now on the right hand but out of sight. A vague track curves left and uphill towards a whitish gate where a little used path heads over the fell to Great Langdale, skirting Blindtarn Moss on the way. Ignore this and walk straight ahead across boggy ground towards a

Exploring Lakeland Waterfalls.

wooded cleft where the gill will soon be heard, then seen. Ascend the true right of the steep, grassy cleft beside the water where its lively performance may be seen through fringing trees. Here are two elongated cascades and some way above, an energetic 3m fall. The source of the gill is of course, Blindtarn Moss, well avoided. It was somewhere there or on the fell above, so the tale is told, that George and Sarah Green, while returning from a visit to Langdale, perished in the great snowstorm of 1808, leaving their six children alone for several days at Blindtarn Cottage.

On the return from the gill a glance up the valley revealed two broad, white cascades, brilliant in the late afternoon sun, against the background of the autumnal colours of upper Easedale. With a fine head of water from recent rainfall, this was the Sourmilk Gill section of the Easedale Beck where it tumbles out of its tarn, impressive indeed after the modest cascades of Blindtarn Gill.

Distance from Grasmere:- 2½k.   *Time:-* 45 mins.

(5) WRAY GILL, GRASMERE

Windermere.   **R. Brathay.**   **NY 328074.**

On a November day with brilliant sunshine after a touch of frost Grasmere was at its best, subtly decorated with autumn tints. The Lake District National Park car park, located in the centre of the village at 338077 made it easy to reach both this and Blindtarn Gill.

To reach Wray Gill walk through the village to the Red Bank (Langdale) Road; after the sharp left hand bend look out for the finger post at 334073 on the right where a gate gives access to a walled lane, canopied by trees. Pass through the gate at the end of the lane and cross the meadow to another gate. Turn right after the short fence to follow a path beside a wall, but the path soon fades among bracken and fallen trees. Cross two tiny gills; persistence pays and ears then eyes perceive the deep-cut channel of Wray Gill ahead. When practical, cross to the true left bank and ascend beside the cascading water. Where the gill becomes enclosed within a

Sector Seven: Falls in the Windermere Catchment.

steep and craggy gully move away, for here danger lurks. Move right or north to less steep ground and follow tracks to the intake wall and cross the stile to the open fellside. Circle back south to the upper reaches of Wray Gill to find more falls then follow tracks which skirt above the crags of Silver How to join the prominent path that leads safely downhill below the crags to the starting point. This is not recommended for children.

Distance for the round:- 4½k. *Time:- 2 hrs.*

(6) SOURMILK GILL, EASEDALE, GRASMERE
Windermere.   **R. Rother.**   **NY 320087.**

Of the three Sourmilk Gills in Lakeland, this is the least steep, the most accessible and the most frequently visited, on account of it lying beside a popular walker's route. When in full flow it is dramatic when seen ahead as the walker approaches across the meadow path beside the Easedale Beck, after leaving the metalled road from Grasmere, especially so when the low-angled sun of an autumn afternoon strikes its lower cascades.

Its water is derived mainly from the outflow of Easedale Tarn, which tumbles not too steeply down its bed of shattered rocks and boulders but in the latter stage of descent it puts on its best performance; no dramatic unsupported falls here, but none the worse for all that. After rain its two cascades are sufficient to obliterate the scream of jet aircraft hurtling over Dunmail Raise and generate a satisfying white fury of foam.

The approach from Grasmere (see Blind Tarn Gill) constitutes a pleasant half-day walk which may be easily extended in this interesting area. Ascend the broad steep path beside the gill on its true right bank, steep at first beside the cascades then easing until the glint of calm water signals the approach to the tarn. Here in the 1930's one could enjoy refreshments from a stone-built hut, the ruins of which may still be traced, a testament to the popularity of this walk for ladies and gentlemen in earlier times.

Exploring Lakeland Waterfalls.

Gill climbers are sometimes known to practice their skills here.
Distance from Grasmere:- 3k.   *Time:- about 1 hour.*

(7) HIGH FALL, RYDAL GILL

Windermere.	R.Rothay.	NY 365067.

Birk Haggs is a woodland area, part of the Rydal Estate which lies athwart the lower reaches of the Rydal Beck above Rydal Hall, now the Carlisle Diocesan Conference and Retreat Centre. According to a notice on the entry gate permission has been agreed with the Estate to allow public access to the grounds north of the hall where several interesting waterfalls may be seen.

From the A591 at Rydal follow the uphill section of the lane that leads past Rydal Church and Rydal Mount and continue northwards along the ensuing track until a sign and a gate in the estate wall is seen on the right. Pass into the access area to follow the path and steps down to a bridge that crosses the beck. Upstream, two sets of cascades may be glimpsed as an introduction to better things to come. Across the bridge turn left and follow the left bank and path to a fine double fall and cascade with its deep plunge pool. Continue to where an exciting 6m ribbon fall at the back of a shallow water-worn recess pours out like a faucet from beneath a chock stone, to splinter on rock and spray into a deep, dark pool. This is High Fall, a splendid sight with a good head of water and much the best on this beck. Further upstream one arrives at a sturdy bridge which carries the Thirlmere water pipeline; from the north parapet note the cascades above which appear less significant after what one has just seen. The way-marks at the end of the bridge indicate a continuation of the path. One arrives shortly at a weir associated with a reservoir chamber where the twin iron pipes would once perhaps have carried water for the village. This is the end of the access area, for way-marks direct the visitor round and back through a young plantation, eventually leading to the entry steps and bridge.

Sector Seven: Falls in the Windermere Catchment.

There is no stated public access along the track beyond Birk Haggs but it would be a gross omission not to mention **Buckstones Jump** (367077), one kilometre further on. At a sharp bend in the Rydal Beck a broad outcrop of grey rock straddles the flow which forces a passage over and through this natural barrier. Prominent at the tail of this barrage is the Jump itself; a 4m fall where the beck squeezes through a V-shaped gap between opposing cheeks of stone at the centre of a sweep of clean grey rock.

Roadside parking is possible beside Rydal Church (not on Sunday) for a small honesty charge. A bus service between Ambleside and Grasmere stops at the lane end, Rydal. At the Grasmere end of Rydal Water there is ample parking on National Trust land, and a footpath from here to Rydal (the Old Coffin Road) provides an agreeable walk between there and Rydal. The rocky ground at Rydal was too shallow for burial there, hence coffins were carried to Grasmere, which is why Wordsworth is interred there although he died at Rydal Mount. A pleasant return is by the south shore of Rydal Water.

Distance, Rydal Church to High Fall:- 1k.   Time:- 15 mins.
Nat. Trust car park via the coffin road to High Fall:- 2k. Time:- 1 hr.

(8) STOCKGHYLL FORCE, AMBLESIDE
Windermere.   R. Rothay.   NY 385046.

How fortunate for Ambleside to be able to boast a magnificent waterfall within walking distance of the town centre - and how remarkable that so few of the thousands of annual visitors take advantage of an evening stroll through Stockghyll Park when the westering sun shines on the face of the fall.

Exploring Lakeland Waterfalls.

Stockghyll gathers its waters from the vale that lies between the two roads, the A592 to Troutbeck and the Kirkstone Road to Ambleside, that descends southwards from the Kirkstone Inn. The Ghyll enters the park on the east side of the town, cascades beneath a wooden footbridge, divides then combines again lower down to form a capital "Y" in shape when seen from ahead. The true left drop of about 6m is the stronger, its companion more of a steep cascade down broken rocks. A pause and they leap again, strike a ledge and recombine to bounce and plunge with thunderous effect into a boiling cauldron of foaming water from where it escapes via a short cascade between moss-lined rock walls and flows away through the park. The water adopts a more sedate progress as it passes around the north of the town, slips unnoticed beneath the A591 to join the River Rothay in Rothay Park and quietly enters Windermere.

Stockghyll Force lies within a short stroll from the center of Ambleside.

To find the fall locate the steep, sharp bend in the town centre which restricts traffic flow by the Salutation Inn and take the steep road signed "To the Falls" by Barclays Bank. Enter the park by the iron gates and walk upstream beside the Ghyll keeping to the true left bank. The best observation point is from the narrow

Sector Seven: Falls in the Windermere Catchment.

rock-flanked approach guarded by iron railings where the falls may be viewed from ahead in their entirety. It is possible to continue upstream, cross the top of the fall by the bridge and return down the true right side, taking great care along the exposed viewing promenade and regain the original path by a lower footbridge.

Distance from town centre:- under 1k.   *Time:- 20-30 mins.*

(9) SKELWITH FORCE

Windermere.   **R. Brathay.**   **NY 342034.**

The River Brathay, the product of the combined waters of Great and Little Langdale, issues from Elterwater and heads towards Windermere to join the River Brathay before merging with the lake. As the river approaches the Skelwith Bridge Hotel it narrows and trips over an abrupt change of level to form a broad apron fall of about 5m into a plunge pool from which it emerges to continue a bouldery and broadening course downstream. A railed metal bridge gives access to a natural rock viewing platform beside the fall, a popular focus point for visitors. Another short bridge links to a second rock bastion a few metres downstream where slippery scrambles to achieve a better view of the noisy cascade may lure the incautious to a ducking. A later visit after heavy rain revealed a totally different aspect of the Force's character. The pressure of floodwater from upstream squeezed the steel-grey surge into the narrows as it approached the fall, now submerged beneath a solid arch of water which merged into a white re-curving ruff of foam, to descend into a maelstrom of spume and spray with a deep, continuous roar that exuded the power of the cataract.

Access is by way of a footpath, part of the Cumbria Way, which follows the left bank upstream from the bridge beside the hotel. Parking is possible in quiet times on the roadside between the hotel and the slate works; that is, on the B5343 Langdale road from which there are points of access to the riverside path. In busy periods the National Trust car park at Silverthwaite is available ½k

Exploring Lakeland Waterfalls.

west along the road. From here cross the road to gain the riverside path and head downstream to find the fall.

A visit to the two falls, Skelwith Force and Colwith Force, *(See Colwith Force)* makes a fine low level circular walk of about nine kilometres. One may begin from either Silverthwaite or Elterwater car parks. Starting say, from Elterwater car park, turn left along the road, past the Youth Hostel then branch right up the hill past the disused quarries, then take the signed field path to Wilson Place. Turn left here along the road then right past Greenbank Cottages. Cross the beck by the footbridge to Strang End then left or east, along the track to High Park, where a path left through the National Trust woodland leads to Colwith Force. Continue from the fall to the road. Cross the road and follow the Cumbria Way footpath across fields to Park Farm and onwards to join the A593 at Skelwith Bridge. Turn left along the Great Langdale road (A593) to Skelwith Force. Continue along the beck-side path and branch off right to the Silverthwaite car park, or go straight on to the Elterwater car park.

This circular walk may be done in either direction or by starting at either car park. It may be shortened by taking the road south all the way from Elterwater to Colwith Bridge but there is no footpath beside this busy, narrow lane.

Distance for circular walk:- 9k.   *Time:- 3 hours plus quality time.*

(10) COLWITH FORCE

Windermere.	Little Langdale.	NY328033.

The Force may be relied on to produce a satisfying spectacle at any time of year but particularly when in spate. The fall sequence begins a short way upstream where a disused weir is sited, once used to supply a head of water to an early, but now disused hydro-electric plant. Below the weir the stream bed drops and the rock walls constrict where a spine of rock projects from the right bank. In the bed, a rock mass swells to obstruct the flow causing

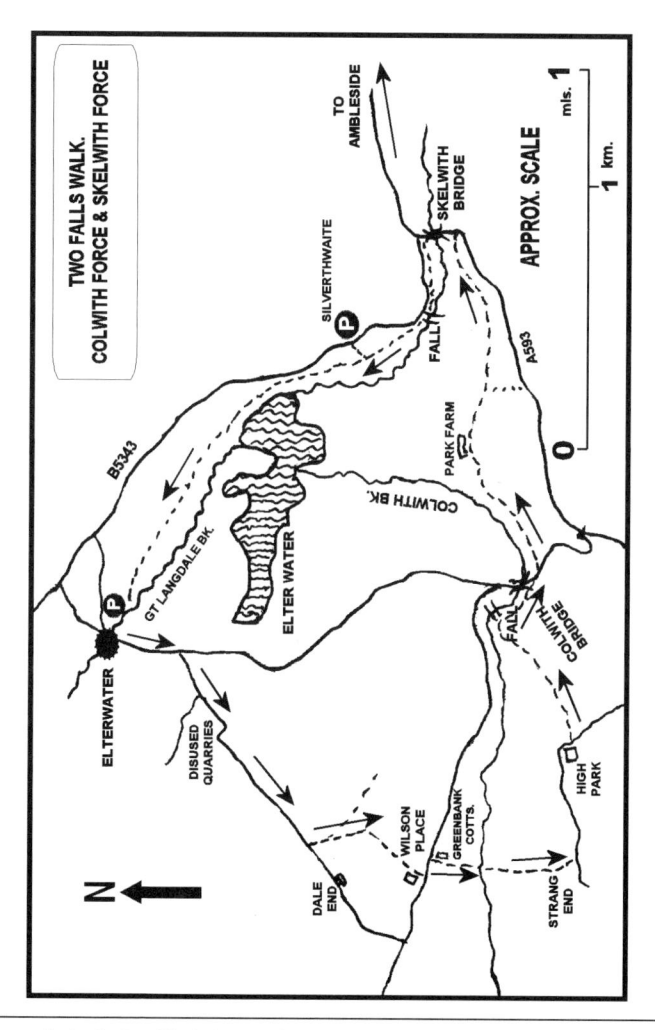

Sector Seven: Falls in the Windermere Catchment.

The walk includes Skelwith and Colwith Forces and takes about 3 hrs.

Exploring Lakeland Waterfalls.

the water to force a passage in five foaming fingers towards the main plunge. The first drop of 4m is split at the lip by a rock projection into two thunderous cascades to be received by a swirling pool from where it is again divided into twin 3m cascades. Its power expended, it flows away north across flat fields to enter Elterwater.

Colwith Force is located on National Trust property close to the road but where there is little parking on the narrow lane. Access to the fall is clearly signed near Colwith Bridge at 330030 where the road from Elterwater Village to the A593 Ambleside to Coniston road crosses the Colwith Beck. Follow the distinct path through the woodland to a ridge which offers a grandstand viewpoint from which to survey the Force.

Colwith Force is five minutes walk from Colwith Bridge.
See also Skelwith Force above.

(11) THRANG CRAG, MEG'S GILL
Windermere.   **Great Langdale.**   **NY 320056 & 324059.**

The falls of significance in the Great Langdale Valley occur on its northern flanks or at its head. Working westwards from Elterwater the above falls occur on the fellside overlooking Chapel Stile, once a thriving centre of stone quarrying; the workings above the village are now disused and activity is concentrated on the opposite side of the valley. Parking in the village is rarely possible, the surest place being Walthwaite Bottom car park at 329051.

Of the two falls, Meg's Gill is the more interesting but both may be combined in a single visit. From Walthwaite Bottom walk up the hill by the minor road and left at the sharp turn for the kilometre road walk to Walthwaite. From the road, ascend the track northwards from the corner of "Walthwaite Lodge" which leads to Meg's Gill. But a short way uphill, take the track left across the disused quarry to Thrang Crag to view the thin, 6m fall. Then return to the main track and continue the ascent into Meg's

Gill. The gully narrows where the gill crosses the path and drops away to the right to form the upper of four stepped falls; the lower ones remain out of sight below, not easy to view from here. Enjoy the exciting traverse around the corner and continue towards the head of the gully where a track swings away to the right to cross Spedding Crag.

Now see the Meg's Gill Falls in their entirety as they boom down the echoing void after heavy rain; first the drops of 3m and 5m crossed on the ascent, then the lower cascades of 7m and 12m separated by a long dogleg of foaming cascades; the whole series constitutes a dramatic sequence of forces. Continue across Spedding Crag above Raven Crag to pick up the path from Grasmere and follow this to descend again to Walthwaite.

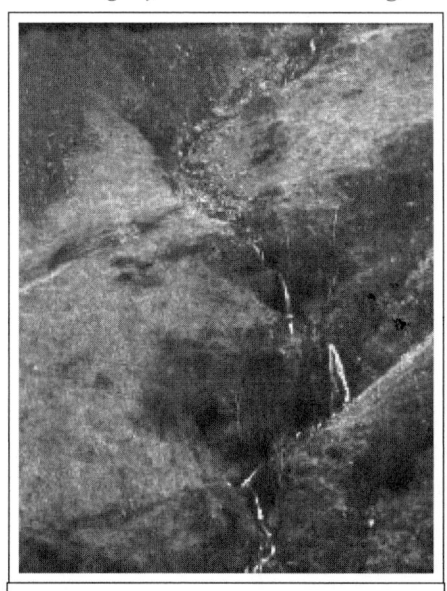

Meg's Gill viewed from Spedding Crag.

Distance; Walthwaite Bottom to top of falls:- 3½k. Time:- 1½ hrs.
Distance; Falls to car park via Spedding Crag:- 3k. Time:- 1½ hrs.

(12) DUNGEON GHYLL FORCE
Windermere.   Great Langdale.   NY 291067.

The main waterfall has much in common with Scale Force, Buttermere, in that it consists of a narrow ribbon of water that lies deep within a dim narrow cleft with sheer or inward-leaning walls.

Exploring Lakeland Waterfalls.

Dungeon Ghyll; within the dim cleft a ribbon of water plunges and a great block hangs above.

In his *"Guide to the Lakes"*, Wordsworth comments that the Force cannot easily be found without a guide, but the modern, sophisticated hill walker may have no difficulty in locating it equipped with today's maps and guide books. Park conveniently at the National Trust car park beside the New Dungeon Ghyll Hotel. Leave the park at its northern end to follow the right bank of Stickle Ghyll upstream and shortly take the path left to the gate on the skyline. Through the gate turn right and in 75m cross the stile to where the Ghyll flows; cross to the left side, or true right of the Ghyll and climb the repaired path. Where the wall bordering the path angles away left, continue on a few paces to where track descends on the right, part way into the Ghyll, to a natural grassy viewing platform beside a sturdy larch. The fall can now be heard and seen, though somewhat obscured by summer leafage.

For the reckless, a slippery descent is possible in the corner where the platform abuts the wall of the ravine where the fall is confined. This leads to further hazards in the form of deep pools

Sector Seven: Falls in the Windermere Catchment.

and slippery boulders which must be negotiated in order to obtain a clear view of this magnificent 12m drop. Forward of the fall, and far above is an oblong, wedged boulder that bridges the cleft. Holly and ash clutch the smooth mossy walls. The floor is littered with great boulders which hold numerous pools within the dim recess where the constant roar of water overwhelms all exterior sound. Return then, to the original path and proceed uphill for 150m to find a series of cascades; in another 70m a broadening of the ghyll reveals a 4m water-slide. Further on, where the gill comes in at a right angle to its lower course, the water tumbles down the fellside in a dramatic 16-18m flurry of foam.

Distance, National Trust car park to main fall:- 750m. Time:- 30mins.
Distance, DG Force to upper fall:- 700m.   Time:- 30 mins.
Please exercise care; this is a dangerous area.

Stickle Ghyll (NY 291070). After visiting Dungeon Ghyll Force one may return to Stickle Ghyll and climb to Stickle Tarn. In the Ghyll itself there are falls a-plenty of modest proportions but the main one occurs at about the halfway stage of the ascent just above where a footbridge carries the track from the right to the left bank where the Ghyll opens out into an amphitheatre with a steep back wall. Here, where trees grow, a steep cascade of about 10m descends to form an eye-catching display dependant upon the amount of surplus water from the dam.

Climb to Stickle Tarn:- 400m ascent from the car park.

(13) WHORNEYSIDE FORCE
Windermere.   **Great Langdale.**   **NY 262054.**

A grand fall this, and a pleasant surprise to come upon it for the first time, tucked away around a corner, just off the main track to Three Tarns and Bowfell. Shy it may be, but this is a robust fall which approaches its descent through Hell Gill and one of my favourites. With its 15 metres or so of total descent and 2m

Exploring Lakeland Waterfalls.

breadth and given average damp Lakeland weather, it may be classed as one of the area's major falls. It consists of three sections, upper and lower are unsupported in full flow while the slightly longer centre section is a steep cascade. A deep, bottle-green circular pool, its saucer-like rim edged with boulders, receives the flow which then passes through two parallel rock grooves to end in a pair of 4m cascades, guarded by a berried rowan.

Whorneyside Force: one of Lakelands most picturesque waterfalls.

A convenient approach is from the Old Dungeon Ghyll Hotel (287062) where there is ample car parking, and along the metalled road to Stool End Farm (277057). Pass between the farm buildings and follow the path until it divides. The right branch leads to The Band, while the left branch, which should be taken, follows the true left bank of the Oxendale Beck. Just past the sheep pen and the first bridge (do not cross) the best path will be found a short distance above the beck; continue to the second bridge which should now be crossed. Instead of continuing on the main path

bear right on the thinner track that follows the beck, now Buscoe Sike, to the fall. Above and beyond the fall is Hell Gill, a dim, grim, narrow, vegetated ravine with a slanting cascade.
Distance, Old Dungeon Ghyll to the Force:- 3½ k.   Time:- 1hr. 15 mins.

(14) BROWNEY GILL
Windermere.   Great Langdale.   NY 265043.

In times of plentiful precipitation Red Tarn - the one situated SW of Pike o' Blisco - sends a steady flow of water down Browney Gill to join the Oxendale Beck. If passing beside the gill alongside Great Knott, either on the way down to Stool End or up towards Crinckle Crags, Cold Pike or associated fells, look out for the interesting 5-6m fall and cascade where it drops into the deep cleft occupied by the combined waters of two energetic left bank tributaries. After descending their fearsome gills in long white water chutes just south of Great Knott, their combined water cascades around a thin rocky arete, where sheltered rowans grow, to meet the Red Tarn outflow. The fall is not recorded on the OS map and may be of interest only in wetter periods.
   Distance from Whorneyside Force:- 2k.   Time:- 30mins.

(15) FALL NEAR HAWKSHEAD HILL
Windermere (via Elterwater).   SD 337985.

This is a pleasant circular walk with good views that begins from the Forestry Car Park at High Cross (333986) and passes close to the fall which is on agricultural land where there may be no public right of way.

From the car park walk downhill along the B5285 road towards Hawkshead. After leaving the village of Hawkshead Hill look out for fingerposts and stiles on the right hand. Take the first stile, (the other leads across fields to Hawkshead) and follow the path beside the watercourse, through the swing gate and across a farm track; the path continues somewhat indistinctly along the same line

Exploring Lakeland Waterfalls.

towards the conifer woodland which is part of the Grizedale Forest Park.

The fall itself is located about 300m to the right along the farm track, beyond a gate in an area known as Frith Plantation. The beck that creates the fall drains from the forestry area of Hawkshead Hall Park from whence it flows northwards until just beyond the gate where it plunges into a deeply scoured gully with precipitous sides, guarded by a stout, stock-proof wire fence. The water descends as a thin slanting chute of 6-7m and flows on to join a similar gully watercourse to become Thurs Gill and past the site of an old bobbin mill.

After crossing the farm track continue along the original footpath, through a gate and across a footbridge, then enter the forest. Take the right hand forest road, then right again at the next crossing to continue in a NNW direction roughly parallel to the forest fence. Finally turn right again until one regains the car park at High Cross.

Distance for the round:- 3½ - 4k. *Time:- 1-1½ hrs.*

(16) RAYRIGG CASCADES
Windermere. **Mill Beck.** **SD 405976.**

This is an interesting water-feature to see on a wet day when, enlivened by spate to enhance its character and conveniently situated close to civilisation, it may be visited with little effort and less discomfort than would be experienced on the fells in adverse conditions.

The Mill Beck rises east of Windermere and flows west through the town to debouch into the lake beside the Steamboat Museum. The interesting section extends between the A5074 and the A592 just north of Bowness. A convenient approach is to start from the lay-by about 200m south of the Steamboat Museum entrance where a rough, narrow lane beside a charming 18th century cottage leads to Longlands (private) Road. Already the beck may be heard

Sector Seven: Falls in the Windermere Catchment.

as it passes through a tunnel beneath the road and behind the lane-side cottages.

Across Longlands Road is signed the "Sheriff's Walk" where a path winds through shrubbery and woodland beside the now visible Mill Beck, to a seat which faces a waterfall of about 4m in height and here drops into the main beck which flows from right to left as one faces the fall. On the occasion of the visit the usually modest fall roared and the gentle cascade of the beck was transformed into a fearsome coffee-brown flood topped with cream, which hissed within its banks.

Continue upstream to where the secret origin of the waterfall is revealed. Here a weir, perhaps man made, causes half the flow to sneak off in a separate channel and disappear into the trees on the opposite bank. A moment's thought and the mystery of the source of the waterfall will be solved! The question remains whether the fall was deliberately contrived to confuse and amuse or simply a freak of nature?

Continue along the waterside path to a stone wall at the corner of which is a worn stone plaque where it is recorded that the Mill Beck Stock Woods were preserved for public use by G.H. Patterson Esquire, High Sheriff of his native county (Cumberland) 1921-22.

Soon, the beck bends out of sight and the path continues to a gate, some cottages and eventually emerges onto the A5074 opposite Dalegarth Hotel at the corner of Queen's Drive, Windermere.

Distance:- under 1k. *Time:- 15 mins.*
Note. If you haven't solved the mystery of the fall, think "island".

Explorng Lakeland Waterfalls.

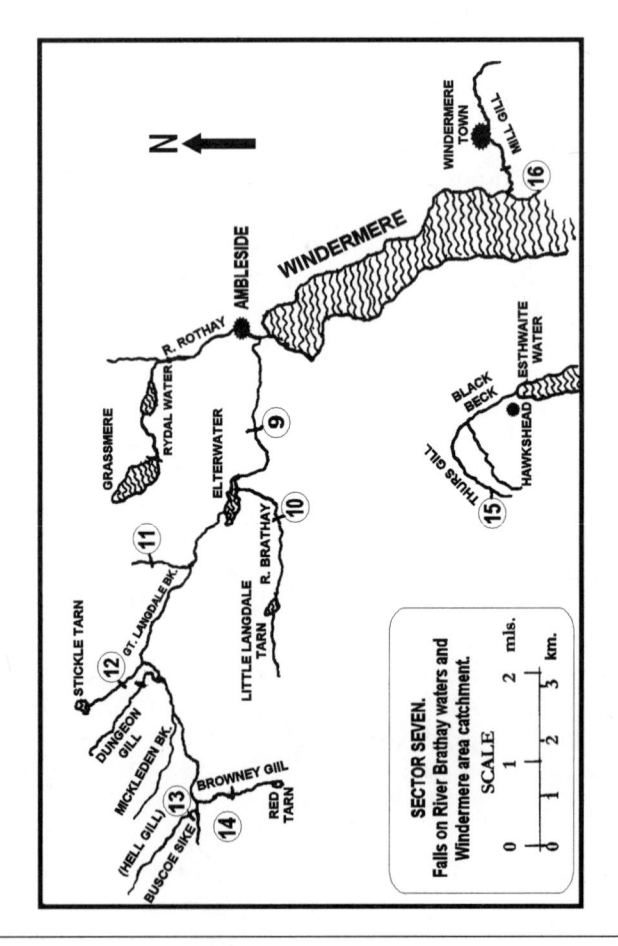

Map to show the location of falls (9) to (16) on R.Brathay streams.

Sector Eight:

Falls in the Coniston Catchment.

Three main streams drain into Coniston from the north and west and on which the main falls in this area occur; these are the Church Beck, the Torver Beck and the Yewdale Beck. Streams on the eastern side of Coniston which flow in from the Grizedale Forest area yield only minor falls and have not been included. Except that is, for the south flowing Grizedale Beck / Force Beck river system which drains the central area of the forest and carries the water towards Morecambe Bay.

Due to the close proximity of the Coniston and Duddon drainage systems both have been conveniently combined on a single sketch map. The fall numbering therefore continues sequentially onwards to the Duddon Falls of Sector Nine. The "West of Duddon" pair lie too far west to be included and are shown on a small separate sketch. So the main falls or fall bearing streams of Sector Eight are as follows:

(1) Church Beck.
(2) Fall below Levers Water.
(3) Fall on Red Dell Beck.
(4) Banishead Quarry Fall.
(5) Ash Gill Beck.
(6) Mill Fall.
(7) Stepping Stones Fall.
(8) Falls in Tilberthwaite Gill.
(9) Falls on Tom Gill.
(10) White Lady Fall.
(11) Force Falls.

The OS maps to cover this area are the Outdoor Leisure 6 and 7.

The sketch map to cover Sectors 8 and 9 is on page 141.

Exploring Lakeland Waterfalls.

FALLS ON CHURCH BECK AND TRIBUTARIES
Coniston. Miners' Bridge, Levers Water, Red Dell Beck.

The area between Coniston Town and Coniston Old Man is a veritable wonderland of hummocky hills, quarry waste tips and mine spoil heaps from the now abandoned copper mines. The Old Man seems to attract plenty of rainfall resulting in the tarns of Low Water and Levers Water, the outflow from these being Low Water Beck and the Levers Water Beck; these combine with Red Dell Beck to form the Church Beck which in turn empties into Coniston Water. To stand in the amphitheatre of "Coppermine Valley" at the confluence of these becks after a period of wet weather is to be surrounded by the echo of continuous rushing water where glimpses of spate-driven cascades may be seen as streaks of white through the hanging mist. There are no famous falls here alas, but as with anywhere in the Lakes, this is a place full of interest and variety.

(1) Fall by Miners' Bridge (SD294982). Starting from Coniston Town there are two ways to approach the first fall at Miners' Bridge, one on either side of the Church Beck. One may start along the metalled road opposite the Coniston Post Office, pass the Museum and circle around Holywath to join the beck on its left bank where the road degrades to a rough track before reaching Miners' Bridge. The advantage of this route is that it provides a clearer view into the beck's deep ravine with its minor falls and cascades.

The right bank approach is made by crossing the bridge in the centre of town on the A593 Torver road and in 20m taking the right turn uphill. Pass through the farmyard and across the fields following the footpath signs. The advantage of this approach is that it provides a better first view of the Miners' Bridge Fall. The fall occurs just downstream of the bridge where a nose of rock in the centre of the stream projects into the ravine and the water

Sector Eight: Falls in the Coniston Catchment.

cascades steeply on either side for about 5-6m. Above the bridge is a 3m cascade where the flow shoots either side of another midstream projection into a rock cleft. Upstream again and just below a weir, water tumbles over a huge jammed boulder.

(2) Fall below Levers Water (SD 283991). Follow now the track on the left bank (that on the right bank will take one to the Old Man) towards the flat delta-like area where miners' cottages, the Youth Hostel and a mining museum are established. Continue by the main uphill track in a north-westerly direction with what is now Levers Beck away on the left hand towards Levers Water and its outflow, just below which is Levers Waterfall of some 6m.

(3) Falls on Red Dell Beck (SD 288989). From the base of the previous fall the remains of a leat or water channel may be followed across the track one has just ascended and around the base of Kennel Crag; this leat once probably carried a reliable water supply to the mine workings in Red Dell, which is the true Coppermines Valley insomuch as this was where most of the mining and processing went on. The leat forms a well-graded track though it later tends to become lost among the clutter of abandoned mine buildings.

Descend to the Red Dell Beck and follow it downstream to cross the wooden sleeper bridge and take the broad track downhill, past the rectangular chimney. Here the beck descends in a series of stepped cascades ending in a fall that cannot clearly be seen from here. Continue down the broad, green track until the ribbon of 5-6m may now be seen. Return to the Youth Hostel and the outward path, perhaps at Miners' Bridge choosing the opposite side of Church Beck for the last leg.

Distance, Coniston to Levers Water:- 3k.   *Time:- 1½hrs.*
Return via Red Dell:- 3½k.   *Time:- 1½hrs.*

Exploring Lakeland Waterfalls.

FALLS ON TORVER BECK
Coniston. **Banishead Quarry.** **Mill Fall.** **Stepping Stones Fall.**

Born of Goats Water in the comb between Dow Crag and Coniston Old Man, the Torver Beck flows south, crosses the Walna Scar Road, flows beneath the A593 Coniston - Furness road near Torver Village and enters Coniston Water at Oxen House Bay. There are several modest falls and interesting features along its course.

The northern section upstream from the A593 bridge may be explored by parking near the sharp bend at 285946 where a minor road strikes north through Crook, bears left then right, past Scar Head and the caravan park to follow the bridleway on the left side (or right bank) of the Torver Beck. On the way there are several small sets of falls, not all accessible to the general public.

(4) Banishead Quarry (SD 277959). A climbing hut is located beside the bridleway and just beyond is a footbridge at 281958 which should be crossed to the opposite side to join the alternative path which has followed the beck's left bank all the way from Little Arrow. Upstream from here, quarry spoil heaps are encountered and 100m from the bridge are the first small accessible falls where water slides down the faces of four steeply inclined slabs of slate, the top cascade at 3m being the largest.

Continue along the track between the spoil heaps noting the confluence of two becks: the one flowing in from the right (left bank tributary) is rectangular in section and obviously man-made. The path follows beside it until it disappears and a stout wire fence takes its place to protect the public from the deep water-filled Banishead Quarry, elliptical in shape and nearly the size of a football pitch. At the SW end a waterfall of about 7m in height pours steadily in, intent on filling the basin but never managing to do so. The curious will question why the basin is never filled to the brim, to overflow and flood the surrounding countryside.

Sector Eight: Falls in the Coniston Catchment.

Torver Quarry and its strange waterfall - and the pool that never fills.

Well, remember that section of beck one has been following -- ? And where, one might ask, does the water for the fall come from? An investigation will show that it is an overflow from the Torver Beck which passes tangentially the far side of the quarry and from which one diverged when the artificial cut was followed. Such are the mysteries of the Banishead Quarries.

Distance, Crook Corner to Banishead Quarry:- 2½k.   Time:- 1½hrs.

For further exploration one may continue northwards to Goat's Water, the source of the Torver Beck or head on for higher things. Alternatively, a gentle return may be made by exploring the falls on **(5) Ash Gill Beck** (SD 273952), provided expectations are modest. Follow the path for 400m north from the water filled quarry to a broken wall on the far side of the beck. Cross and follow the wall SW. The wall marks the boundary of the National Park Access Land of Torver High Common. The green track leads to another area of quarry tips; just beyond is the Ash Gill Beck.

Exploring Lakeland Waterfalls.

Follow downstream past the cascades generated by backward-tilting slates in the narrow stream bed down which the water chuckles merrily. On reaching level ground it wanders away NE to become Tranearth Beck and join the Torver Beck. Cross the Bull Haw Moss Beck by the stepping stones, skirt Wide Close then pass between High and Low Torver Park opting left for Crook Corner or right for the Inn at Torver.

Distance, Banishead Quarry to Torver:- 3½k.   *Time:- 2 hrs.*

(6) Mill Fall (SD 286934). Approximately 400m SW from the

Mill Fall, a cascade on the Torver Beck. A race carries water to a disused waterwheel; see illustration below.

junction of the A593 and the A5084 in Torver Village a signed footpath heads across fields and meets a metalled lane and bridleway. Continue along the lane but where it bears right take the bridleway left that runs beside the tranquil flow of the Torver Beck. Ahead the sound of water indicates the presence of the fall but alas, denied to public view by a secure wire fence.

Sector Eight: Falls in the Coniston Catchment.

The lip of the fall is angled at 45 degrees to the flow; the water descends the stepped, near vertical rock strata in braids, dropping from level to level to create a busy cascade of 5-6m into a sheer-sided, moss-walled channel. A race from the head of the fall carries water towards the original mill, now a private residence, while a branch flows to a derelict breast-shot waterwheel attached to a corrugated iron shed, perhaps once used for wood turning.

Continue along the bridleway, past Mill House where a gated path left across Mill Bridge, leads up to the A5084 beside Emlin Hall. A view of the fall may be glimpsed through the trees from the path.

Distance, Torver to Mill Bridge:- 1½k.   *Time:- 45 mins.*

The old breast-shot wheel on Torver Bk.

(7) Stepping Stones Fall (SD 287926). The last fall of significance before the Torver Beck finally discharges into Coniston Water may easily be overlooked even though it is just 150m downstream from the stepping stones and stout wooden footbridge that gives access to Torver Common; indeed, it may even be glimpsed through the trees from the A5084 Torver to Blawith road.

About half-a-kilometre south of Beckstones Garage there is parking for about five cars on the

Exploring Lakeland Waterfalls.

east side of the road and just opposite the signed path down to the beck and river crossing. Cross the beck by the bridge then in order to follow its right bank downstream it is necessary to leap a small feeder stream and follow an uncertain path through bracken and gorse but which becomes more distinct, as one approaches the broad, shallow gorge where the falls are situated.

From the right side of the little gorge one may look down on the flurry of white water where a two step cascade of 3-4m thunders down. Immediately above the cascade note that a portion of the beck has branched off and circled around a small rocky islet to rejoin the main flow below the fall. Downstream from here the vigour of the beck declines into a series of rapids as it heads towards the grateful lake.

Distance:- under ½k. *Time:- 15 mins.*

FALLS ON YEWDALE BECK
Coniston.   Tilberthwaite Gill, Tom Gill &White Gill.

The Yewdale Beck flows south from Yew Tree Tarn picking up the outflow from Tarn Hows (Tom Gill) Tilberthwaite Gill and White Gill, to skirt the northern fringe of Coniston Town and enter Coniston just north of Church Beck.

(8) Tilberthwaite Gill (NY 303006). This boomerang-shaped area of National Park Access Land (as it appears on the OS map) encompasses the southern flank of Tilberthwaite Gill, a deep rugged gorge through which flow the combined waters of Henfoot Beck and Crook Beck to become the Yewdale Beck within. The northern flank is administered by the National Trust. A track extends along each side of the gorge; that on the south flank rises from the parking area beside the metalled road where a sign gives warning of former quarries nearby; tantalizing glimpses of these may be seen through "windows" beside the path. Soon after ascending the south flank track a green path slants gently down on the right, passes a wooden bench seat and steepens to the bridge

Sector Eight: Falls in the Coniston Catchment.

that crosses the Yewdale Beck. Across the wooden footbridge a steep stone stairway climbs to a stile and beyond to join the north flank path; this is now the only crossing of the gorge. From the bridge one may view the widening gorge downstream where occasionally gill scramblers (school parties mostly) negotiate the bouldery bed. Upstream, the narrowing walls bend to the right where a second bridge, little more than a viewing platform really, once existed. It was built in the late 1960's, but was destroyed by storms during the winter of 1993-94. In Victorian times an iron walkway and viewing gantry was installed here, but long since dismantled. What intriguing feature lurks out of sight around the corner, one wonders?

Not a lot apparently, merely an apron fall of 4-5m; but a feature such as this, in form much like Skelwith Force, can be very impressive in spate. Traces of the approach to the 1960's bridge can still be seen on the north flank in the form of a path, a stile and what may have been a warning notice. The restructured northern path is frequently used as an approach to Wetherlam via Hawk Ridge and Wetherlam Edge, but a circuit of the gorge makes a pleasant stroll in its own right in either direction. The southern path is scrambly and steep in parts and at the head of the gorge where the path levels, two streams may be seen pouring into the hollow which is the start of the Gill. From this viewpoint the Henfoot Beck can be seen as a slim 3m fall. Ford the Crook Beck at the rocks and curve right or north to cross the Henfoot Beck by the wooden footbridge to begin the descent of the north side of the gorge. The Crook Beck can now be seen flowing in by two 3m falls, invisible from the south side.

The path has been well engineered on this side and permits some glimpses into the upper reaches of the gorge where the water seethes and roars in a wall-to-wall cascade. To find the start of the northern path from the road, continue past the parking area and take the signed gravel track by the front of the cottages of Low Tilberthwaite. At the "no public footpath" sign take the swing gate

Exploring Lakeland Waterfalls.

left and continue up the steep path.
Circuit of Tilberthwaite Gill:- 2k.   *Time:- 1-1½ hrs.*

(9) Tom Gill (SD 327999). Out of sight to the east of the A593 Ambleside to Coniston road lies Tarn Hows; the outflow from this picturesque water exits westwards and beneath the road at Glen Mary Bridge (322998) where there is a small parking area at the National Trust Lane Head Coppice. From here, cross the timber bridge and follow the beck upstream on a waymarked track all the way to the tarn. The first fall on the beck, consists of a double cascade which drops 3m then 5m over broken rocks; a pleasing sight in this woodland setting.

Tom Waterfall on the outflow from Tarn Hows.

Continue, keeping on the left side (true right) of the flow; ahead

138

Sector Eight: Falls in the Coniston Catchment.

can be seen and heard the rush of water as one approaches a narrowing ravine where an impressive fall of 5m slides down a steeply inclined rock face. New signs indicate this to be "Tom Waterfall". Above is a staircase of five cascades, the uppermost being some 2 - 2½m. The path edges upwards, easing as the tarn comes into view.

One may now make a circuit of the tarn if desired or return by an alternative track just south of the point where one emerged on the shore of the tarn.

Distance A593 road to the tarn:- under 1k.   *Time:- 20 - 30 mins.*

(10) White Lady Fall (SD 308988). White Gill descends from the Yewdale Fells between Yewdale Crag and Mart Crag and glides down the breast of a smooth 25m buttress to create, in wet weather, a periodic fall of impressive proportions, known as the White Lady, easily seen from the A593 Ambleside to Coniston road which passes below. For the most part it is a trickle or just a dry scar in summer. In a cold winter after rain it becomes a sheet of ice but such conditions are becoming rare.

It may be seen on the westde of the road half way between Coniston Town and Glen Mary Bridge (above). Drivers should encourage their passengers to look out for it rather than do so themselves.

(11) FORCE FALLS.
<u>Windermere.</u>	<u>**Force Mills, Grizedale Forest.**</u>	<u>**SD 340912.**</u>

Force Beck flows southwards from Satterthwaite and through the hamlet of Force Mills, collecting much of its water from the streams that drain from the Grizedale Forest on either side; it eventually joins the River Leven out of Windermere and hence should truly be classified in the Windermere drainage. The beck must deliver a considerable and sustained flow in order to have supported the bobbin mill which once existed here, probably in the barn-like building beside Force Mill Farm.

Exploring Lakeland Waterfalls.

The modest falls are just north of the hamlet and a convenient approach, as there is no parking nearby, is from Blind Lane Picnic area at 344913 on the road NE from Force Mills. From here, walk on prominent paths through the forest, first NW then SW keeping left at each divergence, but outside the walled area, until one emerges on the Satterthwaite road just north of Force Mills. The sound of rushing water as one approaches is a good guide.

The breached reef at Force Mills.

Access to the cascades is possible at several points from the roadside; to be systematic about it one may begin at the small, rusty iron swing gate just north of the top cottage in the hamlet. Walk upstream to view the cascades which are formed by a series of three transverse rock ribs which cross the beck, producing pleasantly noisy white-water cascades with heads of 3-4m. Finally, a fourth natural reef crosses the beck, *(above)* appearing almost weir-like in its formation in which deliberate V-shaped cuts

Sector Eight: Falls in the Coniston Catchment.

appear to have been made; through this the pent-up water roars in a 3m fall. Exploration here might be combined with a visit to the Grizedale Forest Park Visitor's Centre further up the valley.
Distance from Blind Lane Picnic Area to Force Mills:- 1k. Time:- 10 mins.

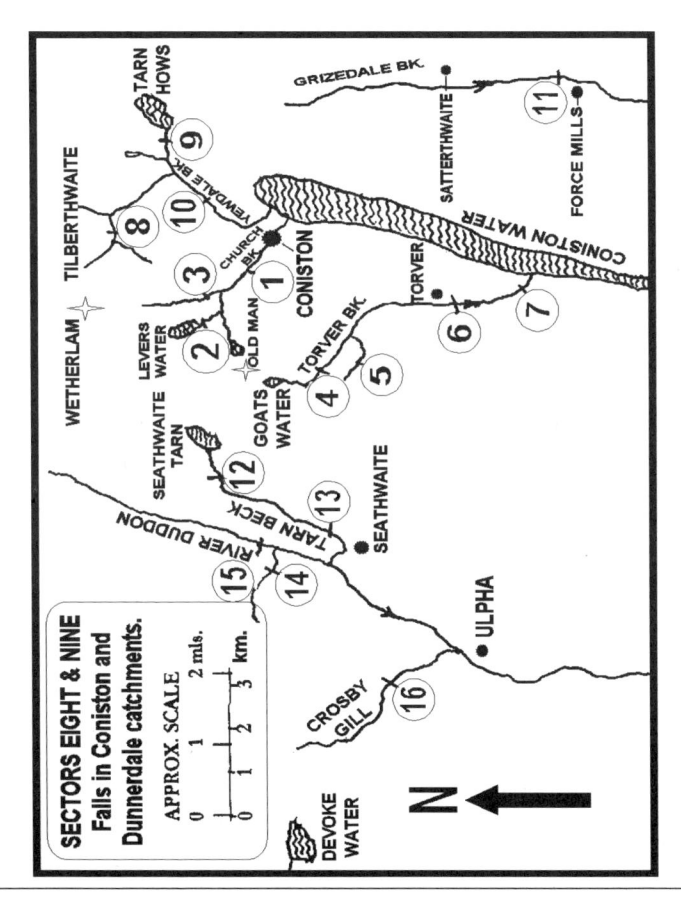

Sketch map to show the location of falls (1) to (11) in sector **8** and falls (12) to (16) in sector **9** below.

Sector Nine:

Dunnerdale.

The Duddon Valley or Dunnerdale is to my mind the most charming of all Lakeland valleys; that of course, is a personal opinion. Less rugged than some yet more rugged than others but with a gentler touch when the sun strikes the golden-brown, weather-smoothed tors that emerge from soft green moorland. Mercifully less accessible and with fewer attractions than the fells to the north of Cockley Beck, it is spared the crowds that throng the more popular crags and still retains something of its character from former days.

The falls in this sector are to be found not on the River Duddon itself but on its tributaries; in addition two other waters are included which lie "West of Dunnerdale" and discharge their water separately into the Irish Sea. These streams and falls are as follows. (The numbering in this sector continues sequentially from Sector Eight - Coniston - with which it shares a map.)

- (12) Falls on Tarn Beck.
- (13) Seathwaite Cataracts.
- (14) Falls on Grassguards Gill.
- (15) Fall on Wet Gill.
- (16) Fall on Crosby Gill.
- (17) Rowentree Force.
- (18) Fall on Buckbarrow Beck.

The OS map to cover this area is the Outdoor Leisure 6.

The sketch map to locate Duddon falls (12) to (16) is on page 141; the one for falls (17) and (18) is on page 148.

Sector Nine: Dunnerdale.

(12) FALLS ON TARN BECK
Duddon Valley. **SD 240985.**

The major fall in the Duddon Valley occurs on Tarn Beck, the outflow from Seathwaite Tarn, enlarged as a reservoir to supply Barrow and District with domestic water supplies. Footpaths in the valley appear to be used infrequently and in consequence tend to become overgrown or obscured by bracken, while waymarking fades as one progresses although initial signposting is clear; this expedition is a case in point.

To find the fall begin from the first cattle grid (234984) south of the Dunnerdale Forestry car park where some verge car parking is possible. A fingerpost on the east side of the road indicates a footpath that rises diagonally up the fellside to a col between Troutal Tongue and High Tongue and offers the first sight of the fall, which from this viewpoint has something of the form of the "Sourmilk Gill" type of waterfall. Continue along the plank walkway in a southerly direction (away from the falls) to a dry-stone wall with stile and continue on a rib of rock into a plantation, to a stone cottage. Pass through two gates then descend diagonally left to a ford and footbridge. Cross to Tongue House Farm (237975), walk through the yard to the gate where waymarks will be seen on either side of the gateway once one has passed through. Take the route northwards or to the left along the walled lane and through the waymarked gate at the end. (This is the last waymark on the walk).

A boggy track is followed to the left hand gate in a wire fence then onwards to a wall with a ladder stile; here Tarn Beck is close on the left hand and here too, is another footbridge, but ignore it. Continue upstream along the valley, through bracken and past an embryo cairn atop a large boulder, through the gap in a broken wall to the base of the falls.

Here the Tarn Beck slants down the fellside for something over 120m, broken and braided by protruding crags into a series of

Exploring Lakeland Waterfalls.

The falls on Tarn Beck, the outflow from Seathwaite reservoir, are of the "sourmilk" type. They emit a low, continuous roar.

cascades and falls. Where the slope increases towards the base of the crags, there occur larger drops of 5-7m, partially enclosed within individual gullies and concealed by overhanging trees. The confusion of sight and sound of descending water gives an impression of magnitude far in excess of the actual flow that finally gathers to form the single stream that passes beneath the footbridge at the bottom.

To return, cross the footbridge here and continue up the valley heading towards the sky-lined power lines and wall with ladder stile. Cross and aim for the latched gate in the next wall then straight ahead to the rocky knoll. Circle right-about the knoll, then left and downhill towards the bottom corner of the field and gate in the supplementary fence that parallels a broken wall. Follow the track and a wall, now on the right hand, past Brow Side Farm, avoiding the gates marked "Private" and keeping the same line, arrive back at the valley road, cattle grid and fingerpost that points back at the way one has just come.

Distance for the round:- 4k. (2½ mls.)   *Time:- 1½ hrs.*

(13) Seathwaite Cataracts (SD231963).

These consist of an exciting set of cascades generated by a series of rocky obstacles that impede the rushing waters of the steeply flowing Tarn Beck, born of Seathwaite Tarn. They are to be found just north of Seathwaite Village and lie within a stone's-throw of the road where there is limited parking in a small lay-by.

Their attraction is in the sight and sound of furiously rushing water leaping a series of rapids and large midstream boulders which frustrate the flow, where athletic photographers (on the occasion of this visit) may clamber up and lie full length to record the unleashed energy of the water transmuted into foam. As the cataracts are so easily accessible they attract a frequent audience of passing visitors during the season and are well worth inspection. Nearby is the Newfield Inn, also worth checking out.

Exploring Lakeland Waterfalls.

(14) FALLS ON GRASSGUARDS GILL
Duddon Valley.   **Dunnerdale Forest.**   **SD 226976.**

Stepping stones are a feature of the Duddon Valley, there being five sets crossing the river between Cockley Beck and Ulpha. Those named Fickle Steps at 228976 provide a crossing from the valley road, where there is lay-by parking for several cars, to the Grassguards track and may be safely negotiated with assistance of a taut, breast-high wire - except in times of spate. Here a sign indicates the alternatives: downstream to Seathwaite, upstream to Birks Bridge or over the hill to Grassguards and eventually Jubilee Bridge at the bottom of Hardnott Pass in Eskdale.

Take the third alternative uphill and in about three minutes note the partially fenced gorge on the left from which the gill emerges via a gentle 3m cascade. Another cascade will be found a further 150m upstream after returning to the main track and proceeding carefully around the edge of a steep crag to a depression covered with fallen beech leaves. Another 100m upstream will bring the main fall into view, a fine 12m water chute which slips down a groove in the smooth, mossy-green rock to merge into a dark plunge pool; in time of spate this becomes a thunderous fall. An introductory cascade of 2½m feeds this feature from above.

Return to the track for a safe descent to the stepping stones; from here the riverside path may be followed upstream to Birks Bridge (not named on the OS map) which is the second crossing point at 235994, for a return to the Dunnerdale Forestry car park. On the way note the fall on Wet Gill which follows below.

Distance:- 2½k.   *Time:- 1 hour.*

(15) Fall on Wet Gill (229978).
After following the River Duddon upstream from the stepping stones for a distance of 300m, one crosses a dry gill and then comes unexpectedly upon the fall of Wet Gill which cascades down a nearly vertical, brooding rock-face of 4½m, fed by a busy

Sector Nine: Dunnerdale.

beck that zigzags through the trees from above. After Grassguards this is safe viewing indeed. (The trees were recently felled, exposing Wet Gill to view from the road.)

By this time the stoney riverside path has become easier with occasional timber walkways. After the first bridged crossing of the river is passed the path detours inland to avoid a steep rock bluff and continues to Birks Bridge, unmistakable with its stone arch firmly anchored to the country rock on either side of the shallow gorge, full of deep pots where the torrent elbows between craggy walls.

(16) FALL ON CROSBY GILL
Duddon Valley. **SD 194948.**

Crosby Gill, which drains the moorland in the Hesk Fell area, passes beneath the Eskdale - Ulpha Moor road at Crosbythwaite Bridge to its rendezvous with the River Duddon at Crosby Bridge, half-a-kilometre NE of Ulpha. Just north of Crosbythwaite Bridge a public footpath extends from 188953 to the Duddon Valley road at Low Wood (206945), and accompanies the gill for a short distance and from which the fall may be seen. Although the section of path at the Duddon end, where it passes through disused quarries, is more interesting, the Crosbythwaite Bridge end is probably more practical in that it provides adequate parking in a lay-by right at the commencement of the path.

Cross the stile beside the fingerpost and head east, but keep on the higher ground to avoid the worst of the mire; there is no distinct path and several drains need to be negotiated. Sag down to the far wall where a gate bears a way-mark; pass through and down to the muddy track beside Crosby Gill. Turn right or south through another gate, past the barn and feeding shelter; the fall is reached moments later and will be seen on the right hand.

The modest fall lies at an angle of about 45 degrees to the flow, is barely 3m in height and 5m in width but partially obscured from

Exploring Lakeland Waterfalls.

view from below by a near-vertical flake of rock, the landward extension of the reef, the probable cause of the feature and a broad, bushy tree. The gill flows placidly above and below this charming fall.

Distance:- under 1k. *Time:- 20 mins.*

<><><><><><>

Falls (17) and (18) located west of the Duddon valley.

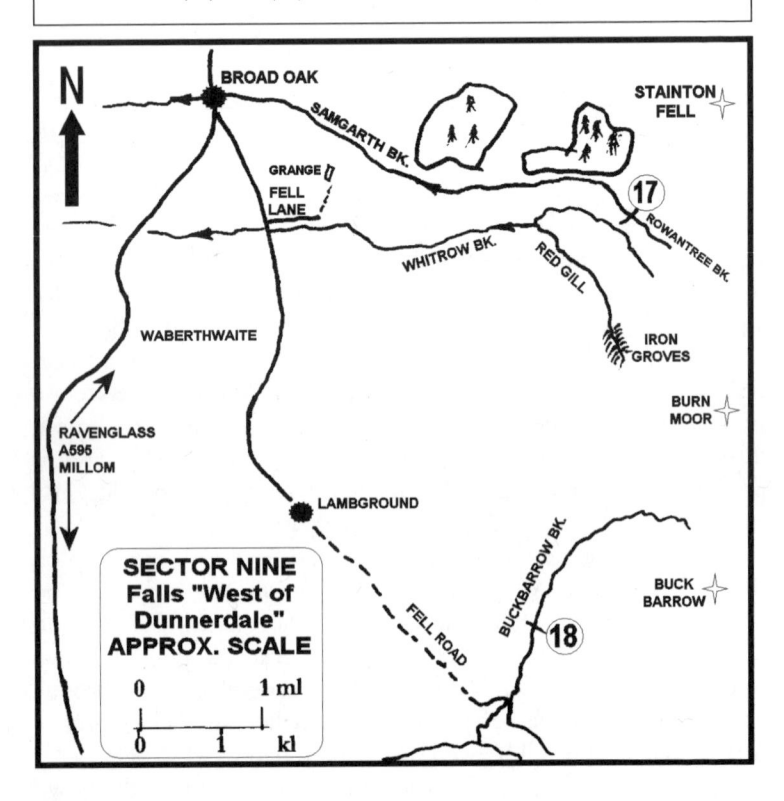

Sector Nine: Dunnerdale.

(17) ROWANTREE FORCE

West of Dunnerdale.　　**Rowantree Gill.**　　**SD 145937.**

West of the Dunnerdale watershed where streams drain to the Irish Sea Coast, the Rowantree Gill descends from the col between Waberthwaite Fell and Stainton Fell where an ancient trackway passes east to west between Ulpha and Waberthwaite - the latter a village famous for its delectable Cumberland Sausage.

A convenient approach to the fall is via Fell Lane (where nearby verge parking is possible at 117938) then walking eastwards along the lane to the gate at the end. There is little indication that the bare fellside ahead conceals a fine fall and attention may stray to the right to the prominent grooves that slash the slopes of Waberthwaite Fell and look as if they might channel a weight of water. They do not. The "Iron Groves" (Grooves?) as they are described on the OS map, are deep and gill-like but channel little water and at the top are virtually dry. Their origin is not immediately apparent, but the rock here appears to contain a high iron content: hence Red Gill.

To reach Rowantree Force avoid the above distraction and head east across the boggy moorland, keeping the conifer plantation on the left hand where the gill is concealed within its deep-cut bed. Where the ground rises a wire fence edges its far bank. As the fellside steepens, so the gill bed deepens until it attains the proportions of a dark, narrow ravine Here, two falls occur one above the other, the lower being a ribbon fall of some 6m, the upper one of 5m. A winter flow made this pair an unexpectedly fine sight. Several leafless rowans edged the sheer drop but caution and the contortions of the cleft denied a clear overview. Upstream, where the gorge shallows is a series of cascades and minor falls, culminating in a picturesque lacy cascade of 6m in a more open setting where a single sturdy rowan stands. From here several possibilities may occur to the walker: the ascent of Whitfell perhaps or a visit to Holehouse Tarn by following the wire fence on to

Exploring Lakeland Waterfalls.

Stainton Pike are two.
Distance, Fell Lane to the Falls:- 3k.   *Time:- 1- 1½hrs.*

(18) FALL ON BUCKBARROW BECK
West of Dunnerdale.   **Corney Fell.**   **SD 137910.**

The modest Buckbarrow Beck occupies a deep furrow in the SW slope of Corney fell, then levels out to traverse a broad, flat-bottomed valley that required the construction of the intriguing Buckbarrow Bridge to carry the Corney Fell Road between Bootle and Muncaster, used frequently as an alternative to the coastal A595. The fall occurs a kilometre north of the fell road which is narrow with passing places; these should not be confused with lay-bys. There are no convenient parking places nearer than Lambground (121916) for the trek to the fall unless the moor is dry, when it may be possible to risk the muddy verge at the start of the path which heads from 130906 or 132904 in a north-easterly direction towards Whit Crags, where the path converges with the beck.

The main fall consists of a steep and energetic cascade of 3-4m that plunges from the lip of a rock slab into a restricted cleft, to be received by a deep pool. After negotiating a second cascade the beck continues with declining energy downstream to pass beneath Buckbarrow Bridge and shortly to merge with the Kinmont Beck and so to the Irish Sea. After viewing the fall it is worthwhile to continue upstream to Littlecell Bottom then circle round to the right onto Buckbarrow itself before returning to the road.

On the left bank of the beck a short way downstream from the falls there is a disused level or mine, reputed to be an old exploratory working for copper; there were once other copper workings on the east side of Whitfell and in the Ulpha region.

Distance from the road to the fall:- 1k.   *Time:- 20mins.*
Add another 1½k for the walk from Lambground.

Sector Ten:

Falls in the Eskdale Valley.

The long, lovely but lake-less valley of Eskdale contains some of the most remote, intriguing and demanding country in Lakeland. The Esk River rises within the very heart of the mountains where the Scafells impose their dominance in height, remoteness and rugged domain. This is not to suggest however, that this sector contains the highest waterfalls in the region but rather that they occur in the most magnificent settings; some are easily accessible but others demand more extended expeditions to view.

Access to the valley by road demands considerable care by those used to city or motorway driving, for road widths are extremely narrow with few easy passing places and very steep approaches; it is rumoured that the prosperity of the valley is based partly upon the recovery and repair of motor vehicles. Entry from the west coast and the A595 is via Gosforth or Ravenglass; from the south from Broughton - in - Furness through Ulpha, or from the east from Ambleside by way of Little Langdale and over Wrynose and Hard Knott Passes. A narrow-gauge steam railway, La'al Ratty, runs a regular passenger service between Ravenglass and Dalegarth Station beside Boot.

The main waterfalls referred to in this sector are as follows.
- (1) Oliver Gill and Whillan Beck - falls on.
- (2) Stanley Force on Stanley Gill.
- (3) Birker Force.
- (4) Scale Force on Scale Beck.
- (5) Falls on the Lincove Beck.
- (6) Esk Gorge and Cam Spout on the upper Esk.

The map to cover this sector is OS Outdoor Leisure 6.

Exploring Lakeland Waterfalls.

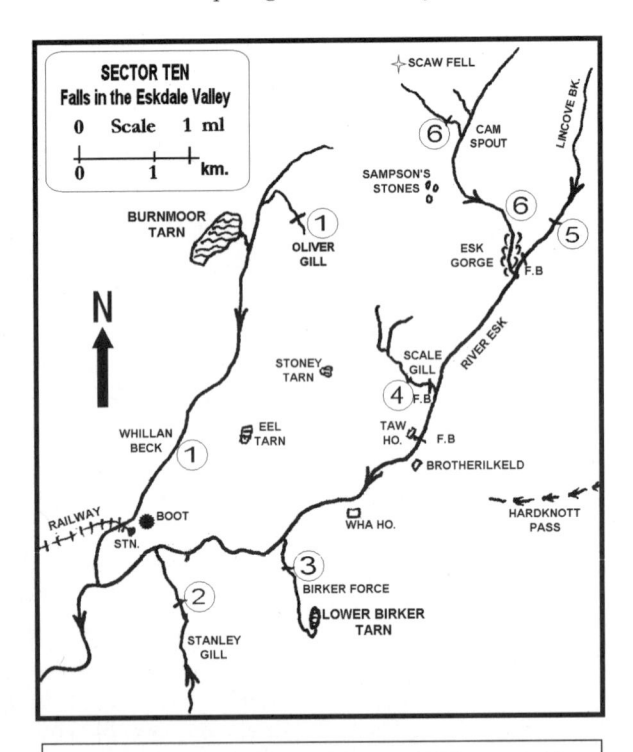

(1) FALLS ON OLIVER GILL

Eskdale.	Whillan Beck.	NY 196048.

A warm June day saw four men, one girl and a dog set off from Dalegarth car park at the La'al Ratty terminus in Eskdale and head north through the village of Boot to follow the Whillan Beck towards Burnmoor Tarn. At the end of the lane the party passed through the gate on the right and onto the bridle path which provided

Sector Ten: Falls in the Eskdale Valley.

> Oliver Gill ; here rest three men, one girl and one large friendly dog. The fall is in summer mode and descends sheer from the crag top right, then diagonally left.

easier ground well above the right bank of the beck; here the cynics and the dog walked, while the fall hunter made for the water's edge.

There are three sets of falls noted on the OS map, the first a short distance beyond the gate; this is a compact 3m. drop into a shallow cutting shaded by oaks. Upstream from here several areas of disturbance reflected unevenness in the stream bed but it was not until the fell gate that the beck became more interesting. Tops of trees appeared round a bend to indicate the presence of a short ravine where the flow rushed and eddied around embedded boulders. Yellow-green oak and polished holly lined the bank: a piece of heaven on this day. Just upstream more cascades occur, then marked by a solitary holly tree is a summer bathing pool fed by a 3m. cascade. At Lambford Bridge quiet meanders ended the first stage of the fall hunt and the warm, dry turf beside Burnmoor Tarn and lunch were irresistible.

Reunited, the party led by the dog, moved on to Bulatt Bridge to cross the outflow from the tarn then leapt, splashed and barked across the marshy bit towards Bleaberry How and Oliver Gill; this has a bonny fall although on this day there was little more than a trickle. The two main falls occur towards the head of the gill, the first a steep cascade of 5-6m down near vertical rocks around which the party dispersed and sat in shirt-sleeve order absorbing sight and sound; potential converts perhaps to the charms of falling water. The upper fall is unsupported, 6m sheer from lip to landing, clinging like a transparent curtain from the top of a buttress that projects above the skyline when viewed from below and from where it would seem no water should be.

The return was brisk over the summit of Great How and Esk Fell, then west skirting Stoney and Eel Tarns to Boot.

Distance for the round:- 11½k. (7 mls.)   *Time:- 4 - 5 hrs.*

(2) STANLEY FORCE

Eskdale.   **Stanley Gill.**   **SD 175995.**

Head up the Eskdale Valley towards Boot. Just before reaching the terminus of La'al Ratty turn right or south opposite Eskdale Village Hall towards Dalegarth Hall where a convenient car park will be found on the left just beyond the bridge that crosses the Esk. Alternatively take the train from Ravenglass.

From the car park walk south in the direction of the Hall where the massive chimneys will be seen through the trees; bear left and follow the signs to the falls. Continue ahead, past a crossing of signed paths towards a gate which advises " To the Falls"; pass through, then bear right with the Stanley Gill on the left hand. The valley narrows and becomes densely wooded, taking on a jungle-like appearance as one approaches the first high level footbridge. Two more bridges follow; at the final one the path divides, one rising to the heights above, the other crossing to the far (true right) side where rock steps and a narrow ledge lead to the viewpoint. A suitable warning on the bridge will deter the infirm and faint-hearted.

Sector Ten: Falls in the Eskdale Valley.

A few paces along the damp ledge, keeping close to the cliff on the left hand, brings the fall into view. In summer volume, a fine veil of water half-a-metre wide drops sheer for 6 metres from lip to pool: a narrow fern-fringed trough between steep mossy walls. Those who decline to cross the bridge and take to the ledge will see only the 3m cascade that empties the plunge pool.

Distance, car park to the fall:- 1k.   Time:- 35 mins.

The first footbridge in Stanley Gill.

Return to the footbridge; if time is not pressing one to catch the next train from Dalegarth Station, take the rising path from the end of the bridge. Where the climb eases, tracks branch right and left. To turn right is to descend to the path back to the car park by Dalegarth Hall without visiting the upper viewpoint which would be a waste of effort. So turn left and up past the sign which reads: "*Warning, 150ft sheer drop. Rock platform viewpoint 100yds.*" A metal railing deters the impetuous from testing the height but an access point permits the cautious to approach the edge.

In the void below, dense vegetation prevents a clear view but it is obviously the gorge one has just left. To the right is the sound of falling water where one may glimpse, through leafy shrubs, its movement. Is this the same fall that one has just viewed from

Exploring Lakeland Waterfalls.

below? That was clearly visible, this is partially concealed in shrubbery. Is this a hidden monster, unseen, unsuspected: the great, undiscovered fall in Lakeland ? Take these mysteries with you as you return, either by the original approach up the valley, or by the right hand path mentioned above.

The adventurous with time on their hands may try the descent on the other side of the valley. Exit the iron fence and walk left the few metres to the stile and open fellside. Follow the fence again left to a swing gate; pass through and cross the bridge over the beck which powers the fall and immediately left yet again, up a steep narrow path through shrubs and past another viewpoint from where one may look back across the gorge to the previous one. Continue along the narrow and little-used path through the bushes to where one's way is barred by a stone wall. Do not descend left. Take the stile right and follow the wall and

Stanley Force. Look above on the right to see where the viewpoint might be.

Sector Ten: Falls in the Eskdale Valley.

vague track downhill; in places the path is steep, slippery and bouldery. *The elderly or infirm should not attempt this descent.* When eventually on the level, spot the ladder-stile on the left and cross. Proceed diagonally right through woodland to the fence with stile and on the same line continue along the path through bracken to a gate. Just beyond is a timber footbridge dedicated to the late Lucy Norris. Beyond the bridge are two paths, one a short distance upstream, either of which will take one back past Dalegarth Hall to the car park.

Distance, viewpoint to car park via east side of gorge:- 2½k. Time:- 1hr.

(3) BIRKER FORCE
Eskdale. **SD 188999.**

Whilst being probably the longest continuous series of cascades in Lakeland, the force is difficult and dangerous to view close up. It extends from high on the south side of the Eskdale valley where the waters from Low Birker Tarn and the moor surrounding it descend from the plateau down steep crags and rucks of huge boulders to the valley bottom to join the Esk. In form it is similar to the "Sour Milk Gill" type of cascade. With a strong head of water to thrust it outwards from the lip, it arcs out and down for 10-12 metres then cascades in a welter of white water for a similar distance down a tight gill.

A pause, then it becomes braided as it negotiates old avalanche debris; as the angle eases the water leapfrogs obstructions and disappears from view within its cleft. Its rush and roar may be heard from across the valley where the road makes an excellent viewpoint to survey most of the fall if binoculars are used. The short upper section, obscured by the cliffs of Gate Crag, must be viewed from above where introductory cascades and a water slide head for the initial arching leap.

From the car park used for visiting Stanley Gill, proceed as if

Exploring Lakeland Waterfalls.

Birker Force: fall and cascades.

for the Gill but at the crossing of footpaths take the one signed "Boot and Upper Eskdale" and follow via a wooden footbridge dedicated to the late Mrs Lucy North which crosses the Stanley Gill to the south bank of the Esk, past a tree-encircled, unnamed tarn, to Lower Birker Farm. Just before the farm, take the grassy track that branches right and zigzags steeply up the valley side until one reaches the roofless remains of a once substantial stone hut. The track just climbed was once the sled-way by which peat from the moor above was brought to the valley. From the hut a narrow but clear track leads to the right towards the head of the Force; take care as you look down the fall and remember the great leap below, which cannot be seen clearly from here. Return down the sled-way and at Low Birker Farm turn right to follow the Esk upstream for a short distance to Doctor Bridge. Cross and return downstream on the other side passing Gill Force, between a wooden footbridge and the stepping stones. At St. Catherine's Church, turn right away from the river, then left along the attractive bridleway to reach the tarmac road. A left turn here will bring one back to the car park.

Distance, car park to Birker Force:- 3.5k.   *Time:- 1½ hrs.*

Sector Ten: Falls in the Eskdale Valley.

Gill Force Fall (NY 177003). This mysterious waterfall prominently indicated on the OS map as occurring on the River Esk has so far eluded me. It may refer to the rippled flow just upstream from the stepping stones near St. Catherine's Church.

Waterfall on Scale Beck downstream from the confluence of the Catcove and the Cowcove Becks.

(4) FALLS ON SCALE BECK
Eskdale. NY 214024.

Take the narrow, walled road up the Eskdale Valley, past Boot, the Woolpack Inn and the Youth Hostel. Opposite Wha House Farm there is an elevated lay-by on the left. In the fence alongside is a stile where the high path or Terrace Route to Scafell begins, sometimes indistinct, past Dawsonground Crags to the Catcove Beck which ripples through cobbley ground. The path continues over Slightside to Scafell but instead, cross the beck to Cat Crag and follow it downstream on the left bank. Beside the crag the beck enters a deep, narrow ravine prominently indicated on the OS map, where the rush of mighty waters may be heard but remain unseen due to its steepness, depth and the trees that crowd its rim. The beck emerges excitedly to join the Cowcove Beck, both unite to become the Scale Beck, one of many of that name in Lakeland. How much more evocative are the names of its tributaries.

Cross the Scale Beck to its left bank, which is easier a short way upstream and follow down to its first fall. This consists of a 2m introductory cascade, immediately followed by a 6m fall into a broad, boulder-bedded gorge. Continue down on the left bank to join the Taw House to Cam Spout trail at the zigzags; the Scale Beck meanwhile, has disappeared but re-emerges from wooded cover lower down as a broad, angry water-slide. The trail shortly crosses Scale Bridge; here is a fine two-stage ribbon fall of 2½ m. then 5-6m: this is Scale Force. It may also be reached more directly by taking the Taw House route from the Eskdale Road starting from Whahouse Bridge or by crossing the Esk by the footbridge from Brotherilkeld Farm (sometimes called Butterilkeld in earlier references) if it is decided to forgo the upstream diversions.

Wha House to Scale Force via Cat Crag:- 3½ k. *Time:- 1½hrs.*
Whahouse Bridge to Scale Force via Taw House:- 2k. *Time:- 1hr.*

(5) FALLS ON THE LINCOVE BECK
Eskdale. **Exploring the Upper Esk.** **NY 228038.**

The most convenient approach is by way of the left or east bank of the Esk starting from the bottom of the Hardknott Pass where there are often parking spaces available, although heavily contested during the summer season, at NY 213013. From here, near the telephone kiosk, a signed path skirts Brotherilkeld Farm and heads north to Lincove Bridge. An alternative but slightly longer approach is to take the path to Taw House from Whahouse Bridge and cross the Esk by the footbridge. The distance to Lincove

Sector Ten: Falls in the Eskdale Valley.

Bridge is 3¼k or (2 miles) of fairly level walking from the telephone kiosk.

The Lincove Beck drains southwards from the Bowfell area to join the R. Esk just below Lincove Bridge. Do not cross the bridge but explore along the left bank, where there are three sets of falls on the beck; the lowest when in spate, manifesting itself as a pillar of white when seen from a distance on the approach. It is fed from above by a series of cascades which themselves seem to coalesce into a similar but shorter pillar. The main fall of 6-7m, plunges into a deep rock pool; the water overflows into yet another pool from where it emerges from the confines of its narrow ravine to flow beneath the bridge and merge with the Esk as its left bank tributary. To reach the second falls ascend the track on the steep slope to follow the descending Lincove upstream still on its left bank to where a series of less dramatic cascades occur. The final display is found just below the point where the Mosedale Valley, which extends SSE. to Cockley Beck, opens on the right hand. Here three energetic cascades are to be found where the water is alternately squeezed between narrowing rock sides or split into separate streams by impeding boulders, to create a display of water gymnastics quite different to the fearsome falls below. The surrounding heights of Crinkle Crags, Bowfell and Esk Pike appear to press down on the lone walker in this unfrequented dale where the sound of water predominates.

Distance to Lincove Bridge:- 3¼ k. (2 mls.)   *Time:- 1 hr.*
Distance, Lincove Bridge to top fall:- 1¾k.   *Time:- 1hr.*

(6) ESK GORGE and CAM SPOUT
Eskdale.   Exploring the Upper Esk.   NY 228036 & 217058.

As Lincove Bridge is approached look across the river to the mouth of the gorge from which the Esk emerges in a series of turbulent cascades and falls. The ravine extends upstream for a kilometre and is flanked by Throstle Crag on its left bank and the notorious Green Crag on the right bank; the passage of the ravine on the right bank is inadvisable, whether ascending or descending and can get one into all kinds of difficulty. At this point anyway, one is on the wrong side of the Esk to attempt it.

Cross the Lincove Beck by the stone bridge and follow the path on the true left bank, above the ravine and beside Throstle Garth where the lower falls step down towards the confluence with the Lincove. Beyond is a deep pool where the gorge widens briefly permitting a foothold for a few isolated trees within, before it narrows to where a double cascade of three, then four metres flows in. As one proceeds alongside the gorge, the path even on this side, becomes more precarious making it advisable to retire to higher ground. The final falls occur beneath Green Crag, not easy to observe with safety although their presence is clearly audible from the refuge of Throstle Crag. There are no monumental falls within the ravine; it is the combination of forces confined within its rock walls rather than any individual fall, that makes the Esk Gorge, especially in spate, such an awesome spectacle. From here the onward scramble eases and the valley widens permitting the Esk and the walker a more leisurely progress, if in contrary directions.

The final objective, the inaptly named Cam Spout, is still 1½k and 30mins. ahead. The easiest approach now is to skirt the western edge of the Great Moss between it and the Esk, using the base of the old deer fence, now much eroded, as a convenient causeway. The fence, long since removed, once formed an enclosure for deer kept by the monks of Furness Abbey; the base was constructed of boulders compacted with peat. Cam Spout Crags rear impressively on the left hand but the Spout is hidden round the corner at the northern end of the crag.

Cross the Esk just above the point where it is joined by the Howe Beck and follow the latter upstream on its left bank. The waterfall of Cam Spout, after what has gone before, may appear disappointing but the setting is magnificent. Here one is surrounded by some of the most spectacular, wild and craggy

Sector Ten: Falls in the Eskdale Valley.

scenery to be found in Lakeland. The fall is but a thin 4-5m feature which degrades into a series of cascades and flows from a knife-cut in the rising ground above that soon expands into an elongated rock basin with the Howe Beck far below its pebbly lip. With the final objective achieved and the day still young, the explorer must decide whether to continue towards Sca Fell Pike or indeed Scafell via the diminutive Foxes Tarn and the repaired path that rises from it. The writer's concern though, is to return the explorer safely to the starting point.

The shortest and easiest route is to follow the How Beck downstream, cross to the right bank and slant southwards beside Cam Spout Crag, past Sampson's Stones (huge boulders like a clutch of monstrous eggs laid by a melting glacier) to the prominent nearby sheepfold. Continue southwards beside the Esk and on the right bank, but where the river swings away eastwards towards Green Crag and the gorge, continue southwards on the well defined track where there is easy walking. Eventually one will descend by the zigzag path to the Scale Beck and Scale Bridge where Scale Force may be inspected in passing. Continue to Taw House and depending on where the transport is waiting, either cross the river by the footbridge to Brotherilkeld or continue on to Wha House Bridge.

Lincove Bridge to Cam Spout:- 3k.   *Time:- 2 hrs.*
Return via Sampson's Stones and Scale Bridge:- 6k.   *Time:- 2½-3hrs*

This is an arduous expedition and should be attempted only in clear weather by experienced walkers.

<><><><><><>

Sector Eleven.

Falls in Wasdale.

In these two west facing valleys there occur fewer falls of interest than one might imagine and then mainly of the "gill" type, of limited catchment and short duration, needing heavy or prolonged precipitation to generate exciting fellside-foaming waterfalls, which any gill will produce under severe conditions This applies particularly to the Wasdale Screes which drain to Wast Water from the narrow Illgill Head - Whin Rigg ridge where streams flow mainly to the south-east into Mitredale.

The land surface to the north and west of Wasdale declines more gently to the Solway Coast, reflected in the long, meandering streams of the River Bleng that joins the River Irt out of Wast Water and which together with the Esk enter the Irish Sea at Ravenglass.

Waterfalls of interest in this Sector, apart from the many-fingered Tongues Gills on Greendale Gill and the little fall and cascade on Ill Gill, both on Seatallan, are not found until the fells increase in height and ruggedness towards the east.

There being so few falls in Ennerdale it was considered convenient to include that valley in this sector.

The streams and falls included here are as follows.

Wasdale: Ennerdale:
(1) Nether Beck -falls on.   (5) Deep Gill- falls on.
(2) Over Beck - falls on.   (6) Low Beck - falls on.
(3) Mosedale Beck - Ritson's Force.
(4) Gable Beck - falls on.

The OS maps to cover these areas are the Outdoor Leisure 6 (for Wasdale) and 4 (for Ennerdale).

Sector Eleven: Falls in Wasdale.

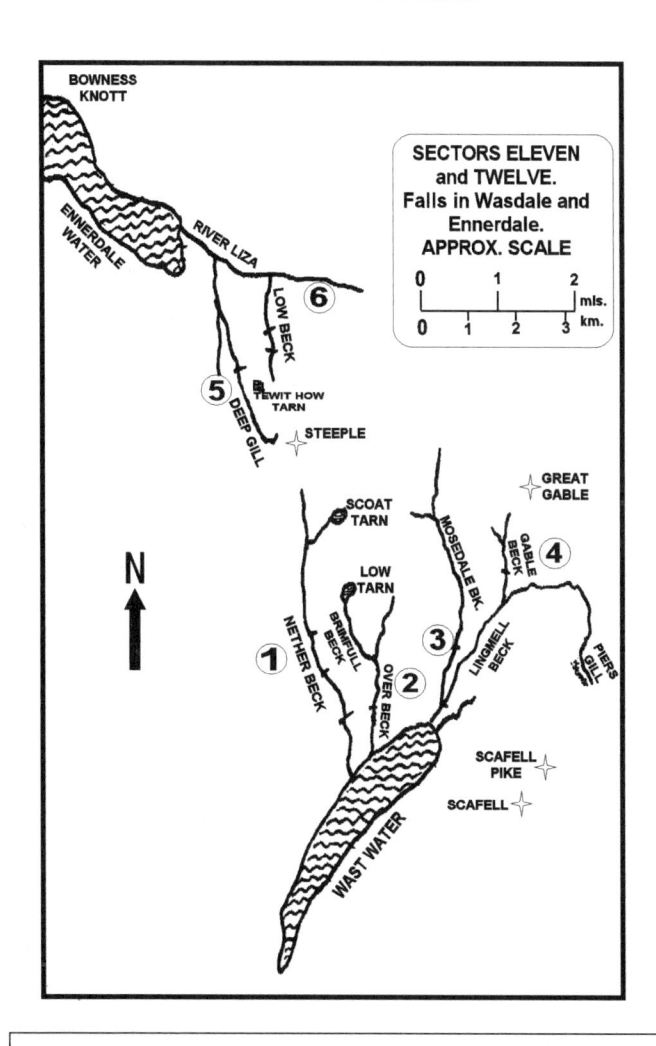

Sketch map to show the location of the main falls in Sector 11, Wasdale: falls (1) to (4) and in Sector 12 Ennerdale: falls (5) and (6), referred to in the text.

Exploring Lakeland Waterfalls..

(1) FALLS ON NETHER BECK

Wasdale. **NY 160072 and NY 160075.**

The first watercourses of interest as one moves up the valley are the Nether Beck and the Over Beck. A convenient car park is situated at Overbeck Bridge where that stream enters Wastwater, from here it is but a short distance back along the road to Netherbeck Bridge and the start of the path that leads upstream beside the beck. This is a particularly beautiful valley and the woodland of mainly oak, holly and rowan which follow the bank of the stream add to its charm.

The first fall, a modest 2m affair, occurs just 150m upstream from the Nether Beck Bridge but soon one reaches the first gorge where, shyly beneath oaks, a silver tongue of water arches outwards to drop 4-5m between sheer rock walls. Two more features of interest occur further on where trees begin to thin; first a broad band of white water that falls into a rounded pool and further yet, where a low rock barrier breaks the flow into three separate channels, two spout out and down, overlapping at right angles as they do so, while the third forms a separate, noisy cascade. After falling 3-4m into a rock-walled pool the beck escapes down a stone channel.

Another but deeper gorge is signalled by a denser growth of trees where Black Beck enters from the left bank to form a thin two-step cascade as it drops to join the Nether Beck. Above this confluence and glimpsed briefly through dense foliage, tumultuous waters roar and foam between jammed, obstructing boulders.

Distance to this point:- 2¾k. *Time:- 1hr.*

From here, another hour's brisk walk upstream, then taking care to cross the beck and follow the next left bank tributary at 154098 (which is flanked by a dry-stone wall and wire fence) will bring one to Scoat Tarn, a delightful, secluded water. From Scoat an easy walk southwards brings one to Low Tarn; on the way as the low ridge is crossed, enjoy the most wonderful panorama across

Sector Eleven: Falls in Wasdale.

Yewbarrow to the Scafells and linking Mickledor: one of the finest vistas in Lakeland.

Follow from Low Tarn the Brimfull Beck until it joins the Over Beck and so complete in one swoop, with no real hard walking, the falls on both becks. (See "Falls on Over Beck" below).
Total distance for the round:- 10k. (6½ mls.) *Time:- 4-5 hrs.*

(2) FALLS ON OVER BECK
Wasdale. **NY 167075 and NY 168086.**

As for the Nether Beck, the starting point may be the car park by Overbeck Bridge (168067). Take the path in the northern corner of the parking place and proceed as if heading for Yewbarrow. Beyond the swing gate continue uphill for 25m to where the fence on the right hand finishes. Follow the indistinct path left, through the gorse to a gate and stile. Cross and follow the now more prominent path through the bracken.

Clear to the ear but partially concealed by leafy oak and berried rowan is a fine 8-10m fall which thunders into a dark and dangerous cleft below the path. The steep and insecure bracken slope ends in a vertical drop that denies clear access to the viewer; this is the major fall on the beck, others of lesser magnitude may be heard and glimpsed through the leafy camouflage as one continues upstream.

The remaining falls lack merit, but soon a bridge spans the rocky trench where cascades roar beneath as the crossing is made to the right bank. Zigzag up beside the wall and turn right along the main track until the rushing Brimfull Beck, which drains Low Tarn, is crossed by a footbridge. Beyond the bridge, rock outcrops beside the Over Beck, below on the right hand, indicate the penultimate cascade of 3½m that fans into a shallow pool after striking a rock pedestal. The final fall is an interesting twin cascade where a pillbox-like outcrop stands sentinel to the upper valley.
Distance:- 2½k. *Time:- 1 - 1½ hrs.*

Exploring Lakeland Waterfalls..

This is a more enjoyable walk if done in the reverse direction and combined with the Nether Beck walk as suggested in that section, the link point being where the Brimfull Beck joins the Over Beck.

(3) RITSON'S FORCE
Wasdale. **Mosedale Beck.** **NY 185094.**

Two main becks flow into the head of Wast Water, the Mosedale Beck and the Lingmell Beck; both combine before entering the lake. The former drains the area between Red Pike and Kirk Fell and it is on the lower reaches of this beck that Ritson's Force is to be found.

Car parking space is available between the head of the lake and the Wasdale Head Hotel except in high season. Walk to the Hotel; to the rear is a slender packhorse bridge, built apparently of vertical slates, which arches over the Mosedale Beck, seemingly too fragile to bear a loaded beast, let alone a couple of walkers. Cross fearlessly and follow the wall to the right to head towards a larch plantation. At the first tree there is a gap in the wall, pass through and head towards the beck where the Force can be heard.

A safe viewpoint beside the beck permits a fine profile aspect of the lower fall, divided by a spine of rock which splits the flow. The right-hand fall as viewed when facing upstream is the larger in volume and height, the drop being about 4m. The white water arcs in a graceful curve, flinging out flecks of spray, then spreads fishtail-wise and links with its sister fall of lesser dimension in the pool below. Some 100m above, an energetic series of cascades and rapids provide an introduction to the Force.

Distance from Wasdale Head Hotel:- ½k.   *Time:- 15- 20mins.*

(4) FALLS ON GABLE BECK
Wasdale. **NY 201097 and 200096.**

Re-cross the packhorse bridge to the Wasdale Head Hotel and

Sector Eleven: Falls in Wasdale.

follow the Mosedale Beck upstream. When the track divides at a small tributary stream bear right, crossing and re-crossing the tributary on wooden footbridges and pass Burnthwaite Farm. At the crossing of Gable Beck, which flows into the Lingmell Beck, head up the steep ascent that would bring one to Beckhead Tarn if followed all the way. Just below where Ill Gill in its deeply indented groove joins the Gable Beck is a fine 8-9m water feature formed by the combined waters cascading down a steep rock wall faced with numerous ledges. In spate the water coalesces to form a fine arching fall; through the descending water the ledge protrusions imprint a watermark pattern of denser whiteness. A few metres below, another 3-4m fall booms into a rock chamber, curls out through a small gap and cascades downhill to the next fall. This is sited just below where a small gill trickles in from the right bank; the fall drops almost sheer for 4-5m between precipitous walls into a deep plunge pool. As with most falls, care is needed in negotiating the steep sides of this beck. In wet conditions the higher of these two falls forms a prominent feature on the fellside and may be seen from the head of Wast Water.

Wasdale Head Hotel to the bottom of Gable Beck:- 1½ k. Time:- 35 mins.

<><><><><><>

Sector Twelve:

Falls in Ennerdale.

(5) FALLS ON DEEP GILL AND (6) LOW BECK
Ennerdale.   **NY 141119, 147115 and 150133.**

Ennerdale extends from east to west and is open to the west coast at that end. It is rimmed by magnificent fells which unfortunately present their more bland profiles to the dale, except for Pillar. Once it was the haunt of fishermen as the names "Char Dub", "Anglers' Crag" and the presence once of the Anglers Hotel on the western shore of the lake, now long since demolished, testify. The shores are given over to forestry and the water is used for domestic and industrial purposes. The only human occupancy is at Gillerthwaite and at the Black Sail Youth Hostel.

The difficulty of entry to the dale emphasises its relative isolation; access is best attained southwards from Cockermouth by the A5086 then through Kirkland or Ennerdale Bridge. Bleach Green at the lake's outflow, the River Ehen, offers parking but in most cases Bowness Knott on the north shore is more practical; there is no motor access beyond these points.

Ennerdale provides excellent walking within the forest, around the lake, or as an access point for some of the higher fells on its rim - but for fall hunters it has little to offer apart from much effort for a small reward.

From Bowness Knott remember there is a 3k or two mile walk each way to Irish Bridge and the crossing of the River Liza before the day begins and ends. From the bridge take the permissive path (indicated by a red, pecked line on the OS map) and follow the true left bank of the Woundale Beck, to note the weir-like cascade just before the confluence of the Silvercove Beck and Deep Gill. Cross these streams by their footbridges and follow the footpath upstream beside the narrowing trench of Deep Gill on its true

Sector Twelve: Falls in Ennerdale.

right bank, which becomes more of a penance than a pleasure, mitigated somewhat by the appearance of some exciting cascades in the stream-way. A further 700m brings relief in the form of a right bank tributary (on the left hand) which permits escape from the confines of this awkward gill if one has not already done so by now. There is even a pleasant fall before one emerges onto the open fell.

Various opportunities are now on offer, one is to follow the tail of the little tributary SE to the fell wall, then left on a broad track to Scoat Fell. Continue to the track northwards onto Steeple, its daunting appearance soon dispelled by a bold, though not reckless approach. The summit achieved, descend now by the north ridge of Long Crag where cairns mark the path, to reach eventually Low Beck and a stile at the forest margin. A series of cascades and falls may now be enjoyed on the direct descent, on a narrow, rough track partly submerged in heather and fern, to the forest road by the lake near Moss Dub.

Should one wish to forego the excitement of Steeple, walk north on emerging from Deep Gill to the diminutive, Tewit How Tarn (146118) now overgrown and follow the course of its outflow until it meets Low Beck at the forest fence. The long trek back to Irish Bridge and Bowness Knott now lies ahead.

Long round via Steeple:- 15k (9½ mls). Ascent:-700m (2100ft). Time:-5-6hrs.

Shorter round via Tewit How 11k (7½ mls). Ascent:- 500m (1500 ft). Time:- 4-4½ hrs.

◇◇◇◇◇◇

Sector Thirteen:

Loweswater and Buttermere.

The main entry to Buttermere is from Cockermouth in the north through the broad Vale of Lorton; the southern end of the valley is curved like a scimitar. Within the valley are the three lakes of Buttermere, Crummock and slightly offset to the west in a shallow side valley; Loweswater. All drain to the River Cocker and northwards to Cockermouth and the River Derwent. On its north-eastern flank there are three passes that link with the Borrowdale Valley: the Whinlatter from Lorton, Newlands from Buttermere Village and Honister, the highest, from Gatesgarth at the head of the valley. The south-western flank is defended by the dramatic Red Pike - High Stile ridge which separates it from Ennerdale. It is at the Red Pike end of this ridge that Scale Force is to be found.

The main falls and fall bearing streams referred to in this sector are as follows.

(1) Holme Force, Loweswater.
(2) Scale Force.
(3) Sourmilk Gill.
(4) Comb Beck - falls on.
(5) Wharnscale Beck.
(6) Spout Force, Whinlatter.

The map to cover this sector os the OS outdoor Leisure 4.

Sector Thirteen: Loweswater and Buttermere.

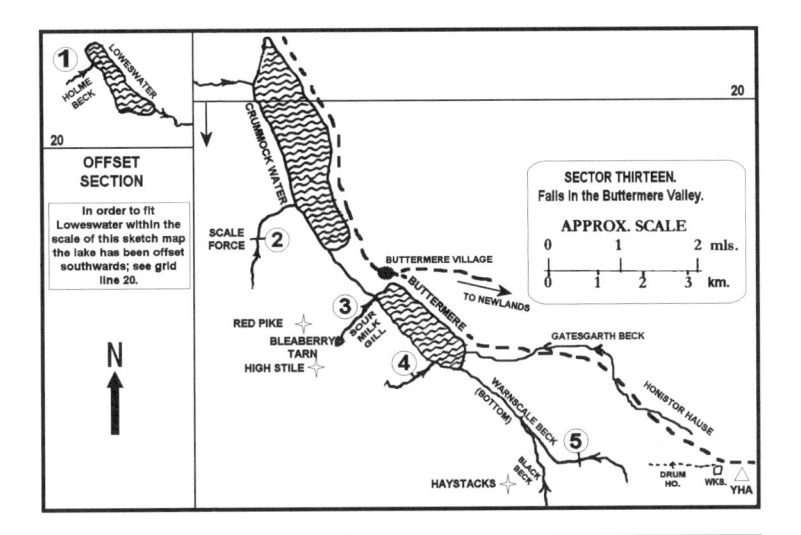

Sketch map to show the location of the falls (1) to (5) in the Buttermere and Loweswater areas. Spout Force is not shown but a clear description of its situation is given it the text.

Exploring Lakeland Waterfalls.

(1) HOLME FORCE

Loweswater Valley. **NY 119214.**

Loweswater is perhaps the least visited of the three lakes of the Buttermere area and for the smaller number of visitors is well supplied with car parking. It provides quiet woodland walks and can offer excellent refreshment in the nearby hostelry of the Kirkstile Inn, adjacent to Loweswater Church. Parking is available on the north shore of the lake at (122224) or near the telephone kiosk at Waterend (118225), both ideal starting points for the walk but do pose a minor logistical problem at the end of the circuit about which advice will be proposed later.

The twin falls of Holme Force

The stopper and spout.

    Walk westwards along the road to take the signed path left just past the telephone kiosk and proceed across the field on duck boards to join the access road at Hudson Place, then bear left along the bridle-path towards the lake and through the gate into Holme Wood.

Sector Thirteen: Loweswater and Buttermere.

Take the first forest track right and uphill until the twin falls are reached; these descend moss covered rocks in the form of a pair of steep water slides about 4m apart which combine within a rocky cauldron to flow below the bridge viewpoint. The division of the Holme Beck occurs above and out of sight, where a unique feature occurs above the point of divergence. This occurs where the water enters a fractured rock trench or ravine; here the water has worn a narrow channel along a ledge that projects from the rock face and slopes at an angle of about 30 degrees down which the beck rushes as a water chute until it hits a stopper, causing it to leap up and over in an arc of spray, to fall into a pool in the lower section of the trench. The ascent to this unusual feature is not recommended as there is no safe path and children should be kept well clear.

After viewing this waterfall there are two alternatives other than retracing one's steps. The shorter is to continue past the fall south-eastwards on the forest track until a path rises from below and cuts diagonally across it. To descend will return one to the main bridle path near the lake and onwards towards Watergate Farm, across fields to Maggie's Bridge and so to the main lakeside road. The problem involves a 2k road walk back to regain one's transport. The solution might be a 1k walk to the local hostelry at Loweswater while a volunteer driver returns to collect the vehicle then enjoys free refreshment, courtesy of his or her grateful companions at the Inn.

If time and energy permit, at the diagonal crossing of paths (above) take the uphill track to the ladder-stile at the woodland boundary. Cross and walk left or SE along the high bridle path to inspect High Nook Tarn then return NE through High Nook Farm to Maggie's Bridge and the final logistical problem.

From car park to fall:- 2k.   *Time:- 30mins.*
From fall to road via Maggies Bridge: 2½k.2   *Time:- 45mins.*
From fall to road via High Nook Farm:- 4k.   *Time:- 1½hrs.*

Exploring Lakeland Waterfalls.

(2) SCALE FORCE
Buttermere Valley.　　　　　　　　　　　　　　　　**NY 151172.**

"Near the head of Crummock Water, on the right, is Scale Force, a waterfall worthy of being visited, both for its own sake, and for the sublime view across the lake, looking back in your ascent towards the chasm. The fall is perpendicular from an immense height, a slender stream faintly illuminating a gloomy fissure". Thus Wordsworth, in his *"Guide to the Lakes"*, describes the fall long recognised as a natural feature attractive to early tourists. Its fame lies in the fact that it is allegedly the highest single drop in Lakeland although estimates of its height vary markedly and range from just under 100 feet to "nearly 160 feet" *(T Bulmer: History, Topography and Directory of Cumberland; 1901.).* It all depends on the point above from which it is measured.

The reasons for its size are puzzling, for the two becks that power it are modest in volume and catchment, nor are the slopes hereabouts excessively steep. Perhaps an existing fault or fracture created a weakness for the descending water to exploit over the millennia since the decline of the last ice age. It is not a beautiful fall, lurking as it does far back in a dim, damp recess just two arms lengths wide, and partially obscured by vegetation that clings to its sheer mossy walls.

To reach the fall, parking is available in Buttermere Village or at the National Trust Car Park a short distance back along the road in the Lorton direction. Walk SW past the Bridge Hotel and the left frontage of the Fish Hotel to take the signed path for Scale Force which branches right to a double-arched stone bridge across Buttermere Dubs, then follow the track right beside Crummock Water, soon bearing left over a shoulder of bouldery moorland. The fall is approached through a gap in a stone wall composed of deep red sandstone-like syenite rock (of the Borrowdale volcanic series) to a footbridge across the Scale Beck.

Walking distance from Buttermere:- 3k.　　　　　*Time:- 50 mins.*

Sector Thirteen: Loweswater and Buttermere.

The "spectre" of Scale Force points to the longest single drop in Lakeland.

(3) SOURMILK GILL
Buttermere Valley. **NY 168158.**

From its source in Bleaberry Tarn, the water of the gill descends the steep slope of 380m to Buttermere in a virtual straight line. When viewed at a distance and in spate it assumes the appearance of a long continuous waterfall of white spray, contrasting with the dark conifer woodland that edges either side of the deeply incised

Exploring Lakeland Waterfalls.

gill. There are no impressive individual falls within its course but a chain of features, from cascades and water chutes to near vertical falls, as the water leaps out from ledges or is split by obstructing boulders.

An earlier path from Buttermere Dubs to Bleaberry Comb once ascended near the edge of the gill, always a dangerous route and it was a deliberate decision on the part of the National Park Authority to construct the present stone path which leads left and upwards to the Comb well clear of the gill, after an accident resulting in the loss of two young lives. Since the construction of the alternative path a more recent attempt by walkers to cross the gill ended in another tragic fatality. It is an extremely dangerous place. So enjoy the gill from the bottom or gaze down from the top near the shore of Bleaberry Tarn - and perhaps ponder how the rising trout within its waters made their way from the lake below against the intimidating force of Sourmilk Gill.

To reach the bottom of the gill and the path to the tarn from Buttermere, walk past the front of the Fish Hotel and onto the field path, then take the left branch to the footbridge over the Buttermere outflow. The red constructed stone path to Bleaberry Tarn is a few metres along on the left, a 380m climb.

On reaching the tarn one may choose to ascend the extra 200m or so to Red Pike summit and descend, with the aid of map and compass, via the path beside Scale Force.

Buttermere to Bleaberry Tarn:- 3½k.　　*Time:- 1½hrs.*
Bleaberry Tarn to Buttermere via Red Pike:- 6k.　　*Time:- 2½ to 3hrs.*

(4) FALL ON COMB BECK
Buttermere Valley.　　Burtness Comb.　　NY 182152.

Comb Beck emerges from Burtness Comb below High Stile and heads downhill for Buttermere. 160m above the lakeside path, from which it is clearly visible, it crosses a short, smooth plateau of grey rock before accelerating over the rounded edge to cascade some 5m over a series of down-sloping ledges in a pretty,

Sector Thirteen: Loweswater and Buttermere.

sparkling display. As the slope levels, it is diverted either side of a small rock islet adorned with velvety grass and surmounted by a single, broad-spreading willow. Further short falls and cascades follow all the way to the lakeside path. In contrast to its intimidating neighbour Sourmilk Gill, this is a more person friendly feature to visit on a summer's day.

This fall on Comb Beck is a delight to visit on a sunny day.

To reach these cascades from Buttermere Village, walk past the left front of the Fish Hotel and take the left branch of the path across the fields to the footbridge which crosses the outflow from Buttermere. Walk along the lakeside permissive path on the edge of Burtness Wood where, just beyond the woodland, the beck will be seen descending from the right.

A return route to Buttermere may be made by completing the circuit of the lake. Continue on from Comb Beck and cross the footbridge and flat land left to Gatesgarth Farm and the valley road. Turn left along the road to pick up the next section of the lakeside path and follow the shore, through the tunnel and on to

Buttermere; a delightful round, with level walking except for the climb to the fall.

Comb Gill from Buttermere:- 2k. *Time:- 50 mins.*
Return to Buittermere via Gatesgarth:- 4½k. *Time:- 1½ hrs.*

(5) FALLS ON WARNSCALE BECK
Buttermere Valley. **NY 205134 and 202135.**

The offer of a lift to Honister Hause by an angler on his way to try his luck in Buttermere was gratefully accepted. A noon clearance after overnight rainfall promised that the becks would be running high and the Wanscale Beck in particular was due for a visit. The old tramway from the quarry works provides an excellent ascent to the high quarries and the ruined Drum House, a junction for paths north to Fleetwith Pike or south to Brandreth and Gable or just over the pass to the quarry track that descends to Wanscale Bottom. Before reaching the Drum House the rain set in again and the descent of slate heaps at Dubs Quarry was well lubricated and needed care before the main downhill track was reached.

The Wanscale Beck which comes in from the left is soon picked up before it begins its plunge into the narrowing defile. The first fall occurs at (205134) in the form of a 2½m wide apron fall of 5m down a steep, knobbly rock face into a swirling plunge pool. A straggly, berried rowan overhung the pool from the left bank and despite the steady, light rain the heather on either flank glowed with rich purple.

Downstream from here things become more serious; from the pool the water surged along a steep water-chute edged with spines of slate and entered a deep, narrow ravine where the roar of confined water was as of a gale of wind in trees. The track here is slippery underfoot with shards of slate on bare rock; it swings right at a sharp corner where the view to Buttermere opens up. The beck meanwhile, freed of the confines of its ravine, hurtled down the open fellside where it split into two main streams either side of

Sector Thirteen: Loweswater and Buttermere.

Warnscale Beck hurtles down between steep banks,
purple-bright with heather.

Exploring Lakeland Waterfalls.

a long, low rock projection, reunited then diverged again in its braided course, producing on each occasion a display of white water reacting to varying degrees of slope and obstruction. A continuous low roar echoed back to the quarry track from the burdened beck. After a final fall of some 6m at 202135 the divided flow recombined, was joined by Black Beck, swept beneath a footbridge and headed along a canalised section for the lake.

A walk along the lake shore, now in bright sunshine, to Buttermere Bridge Hotel for a rendezvous with a happy angler, refreshment and a lift home completed a satisfying afternoon.

Unless a friendly angler can be found, park at Gatesgarth Farm (charges), walk along the road to the signed path right for just over 1k and branch right for the lower falls or continue for 1½k up the quarry track to the upper fall. From just beyond, one may cross Warnscale Beck southwards on to Haystacks and complete the round by returning to Gatesgarth via Scarth Gap.

From Gatesgarth Farm to upper fall:- 3k. *Time:-1hr.*
Return over Haystacks and Scarth Gap:- 4k. *Time:- 2-3 hrs.*

Alternatively, park free at the top of Honister Hause behind the Youth Hostel and follow the walk described above and return back up Honister Pass. Other alternatives may spring to mind.

(6) SPOUT FORCE
Buttermere. Whinlatter Forest (Darling How Plantation) NY182260.

This is a delightful, vertical fall hidden away in the forest, which can be reached within a short distance of convenient car parking from the Whinlatter Pass on the Lorton side. If travelling from the Cockermouth direction on the B5292 the road bends abruptly right then left up a steep gradient and levels out. At this point a signed track off to the left (182255) should be taken for 50m to a car park. Beside the stile ahead there is a diagram of the alternative walks to the fall; inclination or ability will decide.

Sector Thirteen: Loweswater and Buttermere.

The shortest approach to the fall is to cross the stile then follow the grassy path skirting the fence; cross right, over the next stile and continue steeply downhill through the conifer plantation to the Aiken Beck. Cross by the footbridge to where a constructed path leads up to a viewing platform facing the fall. The water upstream is confined within a sheer-sided gorge and drops over a step for 8-9m cascading over projections to become in the lower half, a fine curtain fall into a deep plunge pool; best seen after heavy rain. From the platform, the lower section is partially obscured by a protruding rock.

A return route is possible by following the path upstream until it crosses the beck well above the fall, to join the forest road that may be followed back to the car park.

If approaching from the Keswick direction, pass the Visitors' Centre at the summit of Whinlatter, spot the lone house away to the left, then the narrow lane branching on the left which will give ½k warning of the required right turn to the fall.

An alternative approach to the fall may be made from Scawgill Bridge at the first steep bend of the pass (177257) where there is a lay-by. Walk upstream on the right bank to reach the viewing platform.

From all approaches, distance:- under 1k.   *Time:- 20-30 mins.*

◇◇◇◇◇◇

Exploring Lakeland Waterfalls.

INDEX

FALL, STREAM OR FEATURE.	PAGE.	SECTOR.
Aira Force	72	4
Angletarn Beck	84	4
Ash Gill Beck	133	8
Ashness Falls	17	1
Barrow Falls	18	1
Banishead Quarry	132	8
Birkside Gill	54	3
Blackmoss Pot	22	1
Bleawater Beck	93	5
Beckstones Gill	35	1
Birker Force	151	10
Blindtarn Gill	111	7
Brandy Gill	45	2
Browney Gill	125	7
Buckbarrow Beck	150	9
Buckstones Jump	115	7
Cam Spout	161	10
Caiston Beck	81	4
Chasm, The	73	4
Cascades. The	73	4
Cat Gill	16	1
Caudale Beck	85	4
Charlton Beck	46	2
Church Beck	130	8
Colwith Force	118	7
Comb Beck	178	13
Comb Gill	22	1
Crosby Gill	147	9
Dash Beck	41	2
Deepdale	79	4

Index.

FALL STREAM OR FEATURE.	PAGE.	SECTOR.
Deep Gill	170	12
Dob Gill	64	3
Dockernook Gill	101	6
Dovedale	80	4
Dove Falls	80	4
Dowthwaite Head	74	4
Dungeon Ghyll	121	7
Ellers Beck	32	1
Esk Gorge	161	10
Fisherplace Gill	58	3
Force Crag	33	1
Force Jump	103	8
Force Falls	139	8
Forces. The	79	4
Forces Falls	95	5
Fordingdale Force	91	5
Gable Beck	168	11
Galleny Force	20	1
Gategill Beck	37	1
Gill Force	159	10
Glenderamackin River	47	2
Glenridding Beck	75	4
Grassguards Gill	146	9
Green Burn	109	7
Greenhead Gill	110	1
Greenup Gill	21	1
Grisedale Beck	77	4
Haweswater Beck	92	5
Hawkeshead Hill, Fall on	125	7
Hayeswater Gill	84	4
Helvellyn Gill	55	3
High Fall	114	7

Exploring Lakeland Waterfalls.

FALL, STREAM OR FEATURE.	PAGE.	SECTOR.
High Force	33	1
High Force	73	4
Hobgrumble Gill	96	5
Holme Force	174	13
Hopgill Beck	94	5
Howe Beck	89	5
Kent R., Falls on	103	6
Kilnhow Beck	37	1
Launchy Gill	65	3
Levers Water	131	8
Lincove Beck	160	10
Lodore Force	18	1
Low Beck	170	12
Low Force	33	1
Mart Crag Fall	61	3
Measand Beck	90	5
Meg's Gill	120	7
Mere Gill	68	11
Mill Beck	126	7
Mill Gill	59	3
Mill Fall	134	8
Miner's Bridge	130	8
Miner's Gill	55	3
Mosedale Beck	168	11
Moss Force	28	1
Nether Beck	166	11
Nethermostcove Beck	77	4
Newlands Beck	31	1
Oliver Gill	152	10
Over Beck	167	11
Parting Stone	78	4
Piers Gill	11	intro.

Index.

FALL, STREAM OR	PAGE.	SECTOR.
Raise Beck	53	3
Rayrigg Cascades	126	7
Red Dell Beck	131	8
Red Tarn Beck	76	4
Ritson's Force	168	11
Roughten Gill	43	2
Roughton Gill	44	2
Rowantree Fall	61	3
Rowantree Force	149	9
Rowantreethwaite Beck	94	5
Rowten Beck	76	4
Scale Beck	159	10
Scale Force (Esk)	160	10
Scale Force (Butt)	176	13
Scaleclose Force	27	1
Scalehowe Force	83	4
Scope Beck	29	1
Seathwaite Catts.	145	9
Sidehouse	104	6
Skelwith Force	117	7
Slades Beck	36	1
Smallwater Beck	93	5
Sourmilk Gill (Bor.)	24	1
Sourmilk Gill (Butt.)	177	13
Sourmilk Gill (Gras.)	113	7
Stanley Force	154	10
Stepping Stones Fall	135	8
Southerndale Beck	42	2
Spout Force	182	13
Sprint R.	98	6
Stickle Gill	123	7

Exploring Lakeland Waterfalls.

FALL, STREAM OR FEATURE.	PAGE.	SECTOR.
Stockghyll Force	115	7
Strang End Gill	76	4
Swart Beck	76	4
Swindale Beck	95	5
Tarn Beck	143	9
Taylorgill Force	25	1
Thornthwaite Force	126	7
Thrang Crag	120	7
Thurs Gill	126	7
Tilberthwaite Force	92	5
Tom Gill	138	7
Tongue Gill	108	7
Tottling Stone	66	3
Ulscarf Gill	65	3
Warmscale Beck	180	13
Wet Gill	146	9
Whelpside Gill	54	3
Whillan Gill	152	10
White Lady Fall	139	8
Whitewater Dash	41	2
Whorneyside Force	123	7
Wray Gill	112	7
Wren Gill	98/100	6
Wyth Burn	63	3
Wythop Beck	34	1